Bioinformatics with Python Cookbook

Learn how to use modern Python bioinformatics libraries and applications to do cutting-edge research in computational biology

Tiago Antao

[PACKT] open source
PUBLISHING   community experience distilled

BIRMINGHAM - MUMBAI

Bioinformatics with Python Cookbook

First published: June 2015

Production reference: 1230615

Published by Packt Publishing Ltd.
Livery Place
35 Livery Street
Birmingham B3 2PB, UK.

ISBN 978-1-78217-511-7

www.packtpub.com

Credits

Author
Tiago Antao

Reviewers
Cho-Yi Chen
Giovanni M. Dall'Olio

Commissioning Editor
Nadeem N. Bagban

Acquisition Editor
Kevin Colaco

Content Development Editor
Gaurav Sharma

Technical Editor
Shashank Desai

Copy Editor
Relin Hedly

Project Coordinator
Harshal Ved

Proofreader
Safis Editing

Indexer
Monica Ajmera Mehta

Production Coordinator
Arvindkumar Gupta

Cover Work
Arvindkumar Gupta

About the Author

Tiago Antao is a bioinformatician. He is currently studying the genomics of the mosquito *Anopheles gambiae*, the main vector of malaria. Tiago was originally a computer scientist who crossed over to computational biology with an MSc in bioinformatics from the Faculty of Sciences of the University of Porto, Portugal. He holds a PhD in the spread of drug resistant malaria from the Liverpool School of Tropical Medicine, UK. Tiago is one of the coauthors of Biopython—a major bioinformatics package—written on Python. He has also developed Lositan, a Jython-based selection detection workbench.

In his postdoctoral career, he has worked with human datasets at the University of Cambridge, UK, and with the mosquito whole genome sequence data at the University of Oxford, UK. He is currently working as a Sir Henry Wellcome fellow at the Liverpool School of Tropical Medicine.

I would like to take this opportunity to acknowledge everyone at Packt Publishing, especially Gaurav Sharma, my very patient development editor. The quality of this book owes much to the excellent work of the reviewers who provided outstanding comments. Finally, I would like to thank Ana for all that she endured during the writing of this book.

About the Reviewers

Cho-Yi Chen is an Olympic swimmer, a bioinformatician, and a computational biologist. He majored in computer science and later devoted himself to biomedical research. Cho-Yi Chen received his MS and PhD degrees in bioinformatics, genomics, and systems biology from National Taiwan University. He was a founding member of the Taiwan Society of Evolution and Computational Biology and is now a postdoctoral research fellow at the Department of Biostatistics and Computational Biology at the Dana-Farber Cancer Institute, Harvard University. As an active scientist and a software developer, Cho-Yi Chen strives to advance our understanding of cancer and other human diseases.

Giovanni M. Dall'Olio is a bioinformatician with a background in human population genetics and cancer. He maintains a personal blog on bioinformatics tips and best practices at `http://bioinfoblog.it`. Giovanni was one of the early moderators of Biostar, a Q&A on bioinformatics (`http://biostars.org/`). He is also a Python enthusiast and was a co-organizer of the Barcelona Python Meetup community for many years.

After earning a PhD in human population genetics at the Pompeu Fabra University of Barcelona, he moved to King's College London, where he applies his knowledge and programming skills to the study of cancer genetics. He is also responsible for the maintenance of the Network of Cancer Genes (`http://ncg.kcl.ac.uk/`), a database of system-level properties of genes involved in cancer.

www.PacktPub.com

Support files, eBooks, discount offers, and more

For support files and downloads related to your book, please visit www.PacktPub.com.

Did you know that Packt offers eBook versions of every book published, with PDF and ePub files available? You can upgrade to the eBook version at www.PacktPub.com and as a print book customer, you are entitled to a discount on the eBook copy. Get in touch with us at service@packtpub.com for more details.

At www.PacktPub.com, you can also read a collection of free technical articles, sign up for a range of free newsletters and receive exclusive discounts and offers on Packt books and eBooks.

https://www2.packtpub.com/books/subscription/packtlib

Do you need instant solutions to your IT questions? PacktLib is Packt's online digital book library. Here, you can search, access, and read Packt's entire library of books.

Why Subscribe?

- ▶ Fully searchable across every book published by Packt
- ▶ Copy and paste, print, and bookmark content
- ▶ On demand and accessible via a web browser

Free Access for Packt account holders

If you have an account with Packt at www.PacktPub.com, you can use this to access PacktLib today and view 9 entirely free books. Simply use your login credentials for immediate access.

Table of Contents

Preface   v

Chapter 1: Python and the Surrounding Software Ecology   1
  Introduction   1
  Installing the required software with Anaconda   2
  Installing the required software with Docker   7
  Interfacing with R via rpy2   9
  Performing R magic with IPython   16

Chapter 2: Next-generation Sequencing   19
  Introduction   19
  Accessing GenBank and moving around NCBI databases   20
  Performing basic sequence analysis   25
  Working with modern sequence formats   28
  Working with alignment data   37
  Analyzing data in the variant call format   44
  Studying genome accessibility and filtering SNP data   47

Chapter 3: Working with Genomes   61
  Introduction   61
  Working with high-quality reference genomes   62
  Dealing with low-quality genome references   68
  Traversing genome annotations   73
  Extracting genes from a reference using annotations   76
  Finding orthologues with the Ensembl REST API   80
  Retrieving gene ontology information from Ensembl   83

Chapter 4: Population Genetics   89
  Introduction   89
  Managing datasets with PLINK   91
  Introducing the Genepop format   97

Exploring a dataset with Bio.PopGen   101
Computing F-statistics   107
Performing Principal Components Analysis   113
Investigating population structure with Admixture   118

Chapter 5: Population Genetics Simulation   125
Introduction   125
Introducing forward-time simulations   126
Simulating selection   132
Simulating population structure using island and stepping-stone models   138
Modeling complex demographic scenarios   143
Simulating the coalescent with Biopython and fastsimcoal   149

Chapter 6: Phylogenetics   155
Introduction   155
Preparing the Ebola dataset   156
Aligning genetic and genomic data   162
Comparing sequences   164
Reconstructing phylogenetic trees   170
Playing recursively with trees   174
Visualizing phylogenetic data   179

Chapter 7: Using the Protein Data Bank   187
Introduction   187
Finding a protein in multiple databases   188
Introducing Bio.PDB   192
Extracting more information from a PDB file   197
Computing molecular distances on a PDB file   201
Performing geometric operations   205
Implementing a basic PDB parser   208
Animating with PyMol   212
Parsing mmCIF files using Biopython   220

Chapter 8: Other Topics in Bioinformatics   223
Introduction   223
Accessing the Global Biodiversity Information Facility   224
Geo-referencing GBIF datasets   230
Accessing molecular-interaction databases with PSIQUIC   236
Plotting protein interactions with Cytoscape the hard way   239

Chapter 9: Python for Big Genomics Datasets   247
Introduction   247
Setting the stage for high-performance computing   248
Designing a poor human concurrent executor   254

Performing parallel computing with IPython	**260**
Computing the median in a large dataset	**266**
Optimizing code with Cython and Numba	**271**
Programming with laziness	**275**
Thinking with generators	**278**
Index	**281**

Preface

Whether you are reading this book as a computational biologist or a Python programmer, you will probably relate to the "explosive growth, exciting times" expression. The recent growth of Python is strongly connected with its status as the main programming language for big data. On the other hand, the deluge of data in biology, mostly from genomics and proteomics makes bioinformatics one of the forefront applications of data science. There is a massive need for bioinformaticians to analyze all this data; of course, one of the main tools is Python. We will not only talk about the programming language, but also the whole community and software ecology behind it. When you choose Python to analyze your data, you will also get an extensive set of libraries, ranging from statistical analysis to plotting, parallel programming, machine learning, and bioinformatics. However, when you choose Python, you expect more than this; the community has a tradition of providing good documentation, reliable libraries, and frameworks. It is also friendly and supportive of all its participants.

In this book, we will present practical solutions to modern bioinformatics problems using Python. Our approach will be hands-on, where we will address important topics, such as next-generation sequencing, genomics, population genetics, phylogenetics, and proteomics among others. At this stage, you probably know the language reasonably well and are aware of the basic analysis methods in your field of research. You will dive directly into relevant complex computational biology problems and learn how to tackle them with Python. This is not your first Python book or your first biology lesson; this is where you will find reliable and pragmatic solutions to realistic and complex problems.

What this book covers

Chapter 1, *Python and the Surrounding Software Ecology*, tells you how to set up a modern bioinformatics environment with Python. This chapter discusses how to deploy software using Docker, interface with R, and interact with the IPython Notebook.

Chapter 2, *Next-generation Sequencing*, provides concrete solutions to deal with next-generation sequencing data. This chapter teaches you how to deal with large FASTQ, BAM, and VCF files. It also discusses data filtering.

Chapter 3, *Working with Genomes*, not only deals with high-quality references—such as the human genome—but also discusses how to analyze other low-quality references typical in non-model species. It introduces GFF processing, teaches you how to analyze genomic feature information, and discusses how to use gene ontologies.

Chapter 4, *Population Genetics*, describes how to perform population genetics analysis of empirical datasets. For example, on Python, we will perform Principal Components Analysis, compute F_{ST}, or Structure/Admixture plots.

Chapter 5, *Population Genetics Simulation*, covers simuPOP, an extremely powerful Python-based forward-time population genetics simulator. This chapter shows you how to simulate different selection and demographic regimes. It also briefly discusses the coalescent simulation.

Chapter 6, *Phylogenetics*, uses complete sequences of recently sequenced Ebola viruses to perform real phylogenetic analysis, which includes tree reconstruction and sequence comparisons. This chapter discusses recursive algorithms to process tree-like structures.

Chapter 7, *Using the Protein Data Bank*, focuses on processing PDB files, for example, performing the geometric analysis of proteins. This chapter takes a look at protein visualization.

Chapter 8, *Other Topics in Bioinformatics*, talks about how to analyze data made available by the Global Biodiversity Information Facility (GBIF) and how to use Cytoscape, a powerful platform to visualize complex networks. This chapter also looks at how to work with geo-referenced data and map-based services.

Chapter 9, *Python for Big Genomics Datasets*, discusses high-performance programming techniques necessary to handle big datasets. It briefly discusses cluster usage and code optimization platforms (such as Numba or Cython).

What you need for this book

Modern bioinformatics analysis is normally performed on a Linux server. Most of our recipes will also work on Mac OS X. It will also work on Windows in theory, but this is not recommended. If you do not have a Linux server, you can use a free virtual machine emulator such as VirtualBox to run it on a Windows/Mac computer. An alternative that we explore in the book is to use Docker as a container, which can be used on Windows and Mac via boot2docker.

As modern bioinformatics is a big data discipline, you will need a reasonable amount of memory; at least 4 GB on a native Linux machine, probably 8 GB on a Mac/Windows system, but more would be better. A broadband Internet connection will also be necessary to download the real and hands-on datasets used in the book.

Python is a requirement. All the code will work with version 2, although you are highly encouraged to use version 3 whenever possible. Many free Python libraries will also be required and these will be presented in the book. Biopython, NumPy, SciPy, and Matplotlib are used in almost all chapters. Although the IPython Notebook is not strictly required, it's highly encouraged. Different chapters will also require various bioinformatics tools. All the tools used in the book are freely available and thorough instructions are provided in the relevant chapters of this book.

Who this book is for

If you have intermediate-level knowledge of Python and are well aware of the main research and vocabulary in your bioinformatics topic of interest, this book will help you develop your knowledge further.

Sections

In this book, you will find several headings that appear frequently. To give clear instructions on how to complete a recipe, we use these sections as follows:

Getting ready

This section tells you what to expect in the recipe, and describes how to set up any software or any preliminary settings required for the recipe.

How to do it...

This section contains the steps required to follow the recipe.

There's more...

This section consists of additional information about the recipe in order to make the reader more knowledgeable about the recipe.

See also

This section provides helpful links to other useful information for the recipe.

Conventions

In this book, you will find a number of text styles that distinguish between different kinds of information. Here are some examples of these styles and an explanation of their meaning.

Code words in text, filenames, file extensions, URLs, user input, are shown as follows: "If you are using notebooks, open the `00_Intro/Interfacing_R notebook.ipynb` and just execute the `wget` command on top."

A block of code is set as follows:

```
from collections import OrderedDict
import simuPOP as sp
pop_size = 100
pops = sp.Population(pop_size, loci=[1] * num_loci)
num_loci = 10
init_ops['Freq'] = sp.InitGenotype(freq=[0.5, 0.5])
```

Any command-line input or output is written as follows:

```
conda create -n bioinformatics biopython=1.65 python=3.4
```

New terms and **important words** are shown in bold. Words that you see on the screen, for example, in menus or dialog boxes, appear in the text like this: "The top line is without migration, the middle line with **0.005** migration and the bottom line with **0.1**."

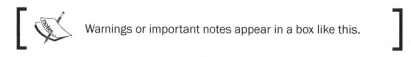

Warnings or important notes appear in a box like this.

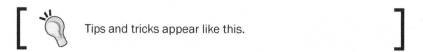

Tips and tricks appear like this.

Reader feedback

Feedback from our readers is always welcome. Let us know what you think about this book— what you liked or disliked. Reader feedback is important for us as it helps us develop titles that you will really get the most out of.

To send us general feedback, simply e-mail `feedback@packtpub.com`, and mention the book's title in the subject of your message.

If there is a topic that you have expertise in and you are interested in either writing or contributing to a book, see our author guide at `www.packtpub.com/authors`.

Customer support

Now that you are the proud owner of a Packt book, we have a number of things to help you to get the most from your purchase.

Downloading the example code

You can download the example code files from your account at `http://www.packtpub.com` for all the Packt Publishing books you have purchased. If you purchased this book elsewhere, you can visit `http://www.packtpub.com/support` and register to have the files e-mailed directly to you.

Downloading the color images of this book

We also provide you with a PDF file that has color images of the screenshots/diagrams used in this book. The color images will help you better understand the changes in the output. You can download this file from: `http://www.packtpub.com/sites/default/files/downloads/5117OS_ColoredImages.pdf`

Errata

Although we have taken every care to ensure the accuracy of our content, mistakes do happen. If you find a mistake in one of our books—maybe a mistake in the text or the code—we would be grateful if you could report this to us. By doing so, you can save other readers from frustration and help us improve subsequent versions of this book. If you find any errata, please report them by visiting `http://www.packtpub.com/submit-errata`, selecting your book, clicking on the **Errata Submission Form** link, and entering the details of your errata. Once your errata are verified, your submission will be accepted and the errata will be uploaded to our website or added to any list of existing errata under the Errata section of that title.

To view the previously submitted errata, go to `https://www.packtpub.com/books/content/support` and enter the name of the book in the search field. The required information will appear under the **Errata** section.

Piracy

Piracy of copyrighted material on the Internet is an ongoing problem across all media. At Packt, we take the protection of our copyright and licenses very seriously. If you come across any illegal copies of our works in any form on the Internet, please provide us with the location address or website name immediately so that we can pursue a remedy.

Please contact us at `copyright@packtpub.com` with a link to the suspected pirated material.

We appreciate your help in protecting our authors and our ability to bring you valuable content.

Questions

If you have a problem with any aspect of this book, you can contact us at `questions@packtpub.com`, and we will do our best to address the problem.

1

Python and the Surrounding Software Ecology

In this chapter, we will cover the following recipes:

- ▸ Installing the required software with Anaconda
- ▸ Installing the required software with Docker
- ▸ Interfacing with R via rpy2
- ▸ Performing R magic with IPython

Introduction

We will start by installing the required software. This will include the Python distribution, some fundamental Python libraries, and external bioinformatics software. Here, we will also be concerned with the world outside Python. In bioinformatics and Big Data, R is also a major player; therefore, you will learn how to interact with it via rpy2 a Python/R bridge. We will also explore the advantages that the IPython framework can give us in order to efficiently interface with R. This chapter will set the stage for all the computational biology that we will perform in the rest of the book.

As different users have different requirements, we will cover two different approaches on how to install the software. One approach is using the Anaconda Python (`http://docs.continuum.io/anaconda/`) distribution and another approach to install the software via Docker (a server virtualization method based on containers sharing the same operating system kernel—`https://www.docker.com/`). We will also provide some help on how to use the standard Python installation tool, pip, if you use the standard Python distribution. If you have a different Python environment that you are comfortable with, feel free to continue using it. If you are using a Windows-based OS, you are strongly encouraged to consider changing your operating system or use Docker via boot2docker.

Installing the required software with Anaconda

Before we get started, we need to install some prerequisite software. The following sections will take you through the software and the steps needed to install them. An alternative way to start is to use the Docker recipe, after which everything will be taken care for you via a Docker container.

If you are already using a different Python version, you are encouraged to continue using your preferred version, although you will have to adapt the following instructions to suit your environment.

Getting ready

Python can be run on top of different environments. For instance, you can use Python inside the JVM (via Jython) or with .NET (with IronPython). However, here, we are concerned not only with Python, but also with the complete software ecology around it; therefore, we will use the standard (CPython) implementation as that the JVM and .NET versions exist mostly to interact with the native libraries of these platforms. A potentially viable alternative will be to use the PyPy implementation of Python (not to be confused with PyPi: the Python Package index).

An important decision is whether to choose the Python 2 or 3. Here, we will support both versions whenever possible, but there are a few issues that you should be aware of. The first issue is if you work with Phylogenetics, you will probably have to go with Python 2 because most existing Python libraries do not support version 3. Secondly, in the short term, Python 2, is generally better supported, but (save for the aforementioned Phylogenetics topic) Python 3 is well covered for computational biology. Finally, if you believe that you are in this for the long run, Python 3 is the place to be. Whatever is your choice, here, we will support both options unless clearly stated otherwise. If you go for Python 2, use 2.7 (or newer if it has been released). With Python 3, use at least 3.4.

If you were starting with Python and bioinformatics, any operating system will work, but here, we are mostly concerned with the intermediate to advanced usage. So, while you can probably use Windows and Mac OS X, most heavy-duty analysis will be done on Linux (probably on a Linux cluster). Next-generation sequencing data analysis and complex machine learning are mostly performed on Linux clusters.

If you are on Windows, you should consider upgrading to Linux for your bioinformatics work because many modern bioinformatics software will not run on Windows. Mac OS X will be fine for almost all analyses, unless you plan to use a computer cluster, which will probably be Linux-based.

If you are on Windows or Mac OS X and do not have easy access to Linux, do not worry. Modern virtualization software (such as VirtualBox and Docker) will come to your rescue, which will allow you to install a virtual Linux on your operating system. If you are working with Windows and decide that you want to go native and not use Anaconda, be careful with your choice of libraries; you are probably safer if you install the 32-bit version for everything (including Python itself).

Remember, if you are on Windows, many tools will be unavailable to you.

 Bioinformatics and data science are moving at breakneck speed; this is not just hype, it's a reality. If you install the default packages of your software framework, be sure not to install old versions. For example, if you are a Debian/Ubuntu Linux user, it's possible that the default matplotlib package of your distribution is too old. In this case, it's advised to either use a recent `conda` or `pip` package instead.

The software developed for this book is available at `https://github.com/tiagoantao/bioinf-python`. To access it, you will need to install Git. Alternatively, you can download the ZIP file that GitHub makes available (however, getting used to Git may be a good idea because lots of scientific computing software are being developed with it).

Before you install the Python stack properly, you will need to install all the external non-Python software that you will be interoperating with. The list will vary from chapter to chapter and all chapter-specific packages will be explained in their respective chapters. Some less common Python libraries may also be referred to in their specific chapters.

If you are not interested on a specific chapter (that is perfectly fine), you can skip the related packages and libraries.

Of course, you will probably have many other bioinformatics applications around—such as bwa or GATK for next-generation sequencing, but we will not discuss these because we do not interact with them directly (although we might interact with their outputs).

You will need to install some development compilers and libraries (all free). On Ubuntu, consider installing the build-essential (`apt-get` it) package, and on Mac, consider Xcode (`https://developer.apple.com/xcode/`).

In the following table, you will find the list of the most important Python software. We strongly recommend the installation of the IPython Notebook (now known as Project Jupyter). While not strictly mandatory, it's becoming a fundamental cornerstone for scientific computing with Python:

Name	Usage	URL	Purpose
IPython	General	`http://ipython.org/`	General
NumPy	General	`http://www.numpy.org/`	Numerical Python
SciPy	General	`http://scipy.org/`	Scientific computing
matplotlib	General	`http://matplotlib.org/`	Visualization
Biopython	General	`http://biopython.org/wiki/Main_Page`	Bioinformatics
PyVCF	NGS	`http://pyvcf.readthedocs.org/en/latest/`	VCF processing
PySAM	NGS	`http://pysam.readthedocs.org/en/latest/`	SAM/BAM processing
simuPOP	Population Genetics	`http://simupop.sourceforge.net/`	Genetics Simulation
DendroPY	Phylogenetics	`http://pythonhosted.org/DendroPy/`	Phylogenetics
scikit-learn	General	`http://scikit-learn.org/stable/`	Machine learning
PyMOL	Proteomics	`http://pymol.org/`	Molecular visualization
rpy2	R integration	`http://rpy.sourceforge.net/`	R interface
pygraphviz	General	`http://pygraphviz.github.io/`	Graph library
Reportlab	General	`http://reportlab.com/`	Visualization
seaborn	General	`http://web.stanford.edu/~mwaskom/software/seaborn/`	Visualization/Stats
Cython	Big Data	`http://cython.org/`	High performance
Numba	Big Data	`http://numba.pydata.org/`	High performance

Note that the list of available software for Python in general and bioinformatics in particular is constantly increasing. For example, we recommend you to keep an eye on projects such as Blaze (data analysis) or Bokeh (visualization).

How to do it...

Here are the steps to perform the installation:

1. Start by downloading the Anaconda distribution from `http://continuum.io/downloads`. You can either choose the Python Version 2 or 3. At this stage, this is not fundamental because Anaconda will let you use the alternative version if you need it. You can accept all the installation defaults, but you may want to make sure that `conda` binaries are in your `PATH` (do not forget to open a new window so that the `PATH` is updated).

    ❏   If you have another Python distribution, but still decide to try Anaconda, be careful with your `PYTHONPATH` and existing Python libraries. It's probably better to unset your `PYTHONPATH`. As much as possible, uninstall all other Python versions and installed Python libraries.

2. Let's go ahead with libraries. We will now create a new `conda` environment called bioinformatics with Biopython 1.65, as shown in the following command:

    ```
    conda create -n bioinformatics biopython biopython=1.65 python=2.7
    ```

    ❏   If you want Python 3 (remember the reduced phylogenetics functionality, but more future proof), run the following command:

    ```
    conda create -n bioinformatics biopython=1.65 python=3.4
    ```

3. Let's activate the environment, as follows:

    ```
    source activate bioinformatics
    ```

4. Also, install the core packages, as follows:

    ```
    conda install scipy matplotlib ipython-notebook binstar pip
    conda install pandas cython numba scikit-learn seaborn
    ```

5. We still need pygraphivz, which is not available on `conda`. Therefore, we need to use pip:

    ```
    pip install pygraphviz
    ```

6. Now, install the Python bioinformatics packages, apart from Biopython (you only need to install those that you plan to use):

    ❏   This is available on `conda`:

    ```
    conda install -c  https://conda.binstar.org/bcbio  pysam
    conda install -c https://conda.binstar.org/simupop simuPOP
    ```

❑ This is available via `pypi`:

```
pip install pyvcf
pip install dendropy
```

7. If you need to interoperate with R, of course, you will need to install it; either download it from the R website at `http://www.r-project.org/` or use the R provided by your operating system distribution.

   ❑ On a recent Debian/Ubuntu Linux distribution, you can just run the following command as root:

   ```
   apt-get r-bioc-biobase r-cran-ggplot2
   ```

   ❑ This will install Bioconductor: the main R suite for bioinformatics and ggplot2—a popular plotting library in R. Of course, this will indirectly take care of installing R.

8. Alternatively, If you are not on Debian/Ubuntu Linux, do not have root, or prefer to install in your home directory, after downloading and installing R manually, run the following command in R:

   ```
   source("http://bioconductor.org/biocLite.R")
   biocLite()
   ```

   ❑ This will install Bioconductor (for detailed instructions, refer to `http://www.bioconductor.org/install/`). To install ggplot2, just run the following command in R:

   ```
   install.packages("ggplot2")
   install.packages("gridExtra")
   ```

9. Finally, you will need to install rpy2, the R-to-Python bridge. Back at the command line, under the `conda` bioinformatics environment, run the following command:

   ```
   pip install rpy2
   ```

There's more...

There is no requirement to use Anaconda; you can easily install all this software on another Python distribution. Make sure that you have `pip` installed and install all `conda` packages with it, instead. You may need to install more compilers (for example, Fortran) and libraries because installation via `pip` will rely on compilation more than `conda`. However, as you also need `pip` for some packages under `conda`, you will need some compilers and C development libraries with `conda`, anyway. If you are on Python 3, you will probably have to perform `pip3` and run Python as `python3` (as `python/pip` will call Python 2 by default on most systems).

In order to isolate your environment, you may want to consider using virtualenv (`http://docs.python-guide.org/en/latest/dev/virtualenvs/`). This allows you to create a bioninformatics environment similar to the one on `conda`.

See also

▸ The Anaconda (`http://docs.continuum.io/anaconda/`) Python distribution is commonly used, especially because of its intelligent package manager: `conda`. Although `conda` was developed by the Python community, it's actually language agnostic.

▸ The software installation and package maintenance was never Python's strongest point (hence, the popularity of `conda` to address this issue). If you want to know the currently recommended installation policies for the standard Python distribution (and avoid old and deprecated alternatives), refer to `https://packaging.python.org/`.

▸ You have probably heard of the IPython Notebook; if not, visit their page at `http://ipython.org/notebook.html`.

Installing the required software with Docker

Docker is the most widely used framework that implements operating system-level virtualization. This technology allows you to have an independent container: a layer that is lighter than a virtual machine, but still allows you to compartmentalize software. This mostly isolates all processes, making it feel like each container is a virtual machine.

Docker works quite well at both extremes of the development spectrum: it's an expedient way to set up the content of this book for learning purposes and may be your platform to deploy your applications in complex environments. This recipe is an alternative to the previous recipe. However, for long-term development environments, something along the lines of the previous recipe is probably your best route, although it can entail a more laborious initial setup.

Getting ready

If you are on Linux, the first thing you have to do is to install Docker. The safest solution is to get the latest version from `https://www.docker.com/`. While your Linux distribution may have a Docker package, it may be too old and buggy (remember the "advancing at breakneck speed" thingy?).

If you are on Windows or Mac, do not despair; boot2docker (`http://boot2docker.io/`) is here to save you. Boot2docker will install VirtualBox and Docker for you, which allows you to run Docker containers in a virtual machine. Note that a fairly recent computer (well, not that recent, as the technology was introduced in 2006) is necessary to run our 64-bit virtual machine. If you have any problems, reboot your machine and make sure that on the BIOS, VT-X or AMD-V is enabled. At the very least, you will need 6 GB of memory, preferably more.

Note that this will require a very large download from the Internet, so be sure that you have a big network pipe. Also, be ready to wait for a long time.

How to do it...

These are the steps to be followed:

1. Use the following command on the Linux shell or in boot2docker:

    ```
    docker build -t bio
    https://raw.githubusercontent.com/tiagoantao/bioinf-
    python/master/docker/2/Dockerfile
    ```

    - ❑ If you want the Python 3 version, replace the 2 with 3 versions on the URL. After a fairly long wait, all should be ready.
    - ❑ Note that on Linux, you will either require to have root privileges or be added to the Docker Unix group.

2. Now, you are ready to run the container, as follows:

    ```
    docker run -ti -p 9875:9875 -v YOUR_DIRECTORY:/data bio
    ```

3. Replace `YOUR_DIRECTORY` with a directory on your operating system. This will be shared between your host operating system and the Docker container. `YOUR_DIRECTORY` will be seen in the container on /data and vice versa.

    - ❑ The `-p 9875:9875` will expose the container TCP port `9875` on the host computer port `9875`.

4. If you are using `boot2docker`, the final configuration step will be to run the following command in the command line of your operating system, not in boot2docker:

    ```
    VBoxManage controlvm boot2docker-vm natpf1
    "name,tcp,127.0.0.1,9875,,9875"
    ```

> On Windows, this binary will probably be in `C:\Program Files\Oracle\VirtualBox`.
>
> On a native Docker installation, you do not need to do anything.

5. If you now start your browser pointing at `http://localhost:9875`, you should be able to get the IPython Notebook server running. Just choose the Welcome notebook to start!

See also

▸ Docker is the most widely used containerization software and has seen enormous growth in usage in recent times. You can read more about it at `https://www.docker.com/`.

▸ You will find a paper on arXiv, which introduces Docker with a focus on reproducible research at `http://arxiv.org/abs/1410.0846`.

Interfacing with R via rpy2

If there is some functionality that you need and cannot find it in a Python library, your first port of call is to check whether it's implemented in R. For statistical methods, R is still the most complete framework; moreover, some bioinformatics functionalities are also only available in R, most probably offered as a package belonging to the Bioconductor project.

The rpy2 provides provides a declarative interface from Python to R. As you will see, you will be able to write very elegant Python code to perform the interfacing process.

In order to show the interface (and try out one of the most common R data structures, the data frame, and one of the most popular R libraries: ggplot2), we will download its metadata from the Human 1000 genomes project (`http://www.1000genomes.org/`). As this is not a book on R, we do want to provide any interesting and functional examples.

Getting ready

You will need to get the metadata file from the 1000 genomes sequence index. Please check `https://github.com/tiagoantao/bioinf-python/blob/master/notebooks/Datasets.ipynb` and download the `sequence.index` file. If you are using notebooks, open the `00_Intro/Interfacing_R notebook.ipynb` and just execute the `wget` command on top.

This file has information about all FASTQ files in the project (we will use data from the Human 1000 genomes project in the chapters to come). This includes the FASTQ file, the sample ID, and the population of origin and important statistical information per lane, such as the number of reads and number of DNA bases read.

How to do it...

Take a look at the following steps:

1. We start by importing rpy2 and reading the file, using the `read_delim` R function:

```
import rpy2.robjects as robjects
read_delim = robjects.r('read.delim')
seq_data = read_delim('sequence.index', header=True,
    stringsAsFactors=False)
#In R:
#   seq.data <- read.delim('sequence.index', header=TRUE,
#   stringsAsFactors=FALSE)
```

   ❑ The first thing that we do after importing is accessing the `read.delim` R function that allows you to read files.

   ❑ Note that the R language specification allows you to put dots in the names of objects. Therefore, we have to convert a function name to `read_delim`.

2. Then, we call the function proper; note the following highly declarative features. First, most atomic objects—such as strings—can be passed without conversion. Second, argument names are converted seamlessly (barring the dot issue). Finally, objects are available in the Python namespace (but objects are actually not available in the R namespace; more about this later). For reference, I have included the corresponding R code. I hope it's clear that it's an easy conversion.

   ❑ The `seq_data` object is a data frame. If you know basic R or the Python pandas library, you are probably aware of this type of data structure; if not, then this is essentially a table: a sequence of rows where each column has the same type. Let's perform a basic inspection of this data frame as follows:

```
print('This dataframe has %d columns and %d rows' %
    (seq_data.ncol, seq_data.nrow))
print(seq_data.colnames)
#In R:
#   print(colnames(seq.data))
#   print(nrow(seq.data))
#   print(ncol(seq.data))
```

   ❑ Again, note the code similarity. You can even mix styles using the following code:

```
my_cols = robjects.r.ncol(seq_data)
print(my_cols)
```

> ❏ You can call R functions directly; in this case, we will call `ncol` if they do not have dots in their name; however, be careful. This will display an output, not 26 (the number of columns), but `[26]` which is a vector composed of the element 26. This is because by default, most operations in R return vectors. If you want the number of columns, you have to perform `my_cols[0]`. Also, talking about pitfalls, note that R array indexing starts with 1, whereas Python starts with 0.

3. Now, we need to perform some data cleanup. For example, some columns should be interpreted as numbers, but they are read as strings:

```
as_integer = robjects.r('as.integer')
match = robjects.r.match
my_col = match('BASE_COUNT', seq_data.colnames)[0]
print(seq_data[my_col - 1][:3])
seq_data[my_col - 1] = as_integer(seq_data[my_col - 1])
print(seq_data[my_col - 1][:3])
```

> ❏ The match function is somewhat similar to the index method in Python lists. As expected, it returns a vector so that we can extract the 0 element. It's also 1-indexed, so we subtract one when working on Python. The `as_integer` function will convert a column to integers. The first print will show strings (values surrounded by "), whereas the second print will show numbers.

4. We will need to massage this table a bit more; details can be found on a notebook, but here we will finalize with getting the data frame to R (remember that while it's an R object, it's actually visible on the Python namespace only):

```
robjects.r.assign('seq.data', seq_data)
```

> ❏ This will create a variable in the R namespace called `seq.data` with the content of the data frame from the Python namespace. Note that after this operation, both objects will be independent (if you change one, it will not be reflected on the other).

> While you can perform plotting on Python, R has default built-in plotting functionalities (which we will ignore here). It also has a library called ggplot2 that implements the Grammar of Graphics (a declarative language to specify statistical charts).

5. We will finalize our R integration example with a plot using ggplot2. This is particularly interesting, not only because you may encounter R code using ggplot2, but also because the drawing paradigm behind the Grammar of Graphics is really revolutionary and may be an alternative that you may want to consider instead of the more standard plotting libraries, such as matplotlib ggplot2 is so pervasive that rpy2 provides a Python interface to it:

```
import rpy2.robjects.lib.ggplot2 as ggplot2
```

6. With regards to our concrete example based on the Human 1000 genomes project, we will first plot a histogram with the distribution of center names, where all sequencing lanes were generated. The first thing that we need to do is to output the chart to a PNG file. We call the R png() function as follows:

```
robjects.r.png('out.png')
```

7. We will now use ggplot to create a chart, as shown in the following command:

```
from rpy2.robjects.functions import SignatureTranslatedFunction
ggplot2.theme = SignatureTranslatedFunction(ggplot2.theme,
            init_prm_translate={'axis_text_x': 'axis.text.x'})
bar = ggplot2.ggplot(seq_data) + ggplot2.geom_bar() +
    ggplot2.aes_string(x='CENTER_NAME')    +
    ggplot2.theme(axis_text_x=ggplot2.element_text(angle=90,
        hjust=1))
bar.plot()
dev_off = robjects.r('dev.off')
dev_off()
```

   ❏ The second line is a bit uninteresting, but is an important boilerplate code. One of the R functions that we will call has a parameter with a dot in its name. As Python function calls cannot have this, we map the axis.text.x R parameter name to the axis_text_x Python name in the function theme. We monkey patch it (that is, we replace ggplot2.theme with a patched version of itself).

8. We then draw the chart itself. Note the declarative nature of ggplot2 as we add features to the chart. First, we specify the seq_data data frame, then we will use a histogram bar plot called geom_bar, followed by annotating the X variable (CENTER_NAME).

9. Finally, we rotate the text of the x axis by changing the theme.

   ❏ We finalize by closing the R printing device. If you are in an IPython console, you will want to visualize the PNG image as follows:

```
from IPython.display import Image
Image(filename='out.png')
```

❑ This chart produced is as follows:

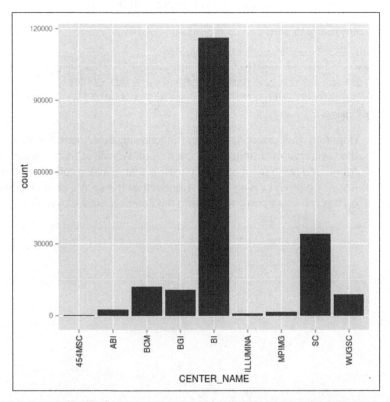

Figure 1: The ggplot2-generated histogram of center names responsible for sequencing lanes of human genomic data of the 1000 genomes project

10. As a final example, we will now do a scatter plot of read and base counts for all the sequenced lanes for Yoruban (YRI) and Utah residents with ancestry from Northern and Western Europe (CEU) of the Human 1000 genomes project (the summary of the data of this project, which we will use thoroughly, can be seen in the *Working with modern sequence formats* recipe in *Chapter 2, Next-generation Sequencing*). We are also interested in the difference among the different types of sequencing (exome, high, and low coverage). We first generate a data frame with just YRI and CEU lanes and limit the maximum base and read counts:

```
robjects.r('yri_ceu <- seq.data[seq.data$POPULATION %in%
c("YRI", "CEU") & seq.data$BASE_COUNT < 2E9 &
seq.data$READ_COUNT < 3E7, ]')
robjects.r('yri_ceu$POPULATION <-
as.factor(yri_ceu$POPULATION)')
robjects.r('yri_ceu$ANALYSIS_GROUP <-
as.factor(yri_ceu$ANALYSIS_GROUP)')
```

- ❏ The last two lines convert the `POPULATION` and `ANALYSIS_GROUPS` to factors, a concept similar to categorical data.

11. We are now ready to plot:

```
yri_ceu = robjects.r('yri_ceu')
scatter = ggplot2.ggplot(yri_ceu) + ggplot2.geom_point() + \
ggplot2.aes_string(x='BASE_COUNT', y='READ_COUNT',
shape='factor(POPULATION)', col='factor(ANALYSIS_GROUP)')
scatter.plot()
```

- ❏ Hopefully, this example (refer to the following screenshot) makes the power of the Grammar of Graphics approach clear. We will start by declaring the data frame and the type of chart in use (the scatter plot implemented by `geom_ point`). Note how easy it is to express that the shape of each point depends on the `POPULATION` variable and the color on the `ANALYSIS_GROUP`:

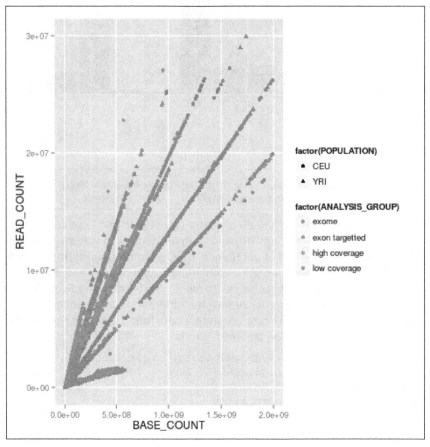

Figure 2: The ggplot2-generated scatter plot with base and read counts for all sequencing lanes read; the color and shape of each dot reflects categorical data (population and the type of data sequenced)

12. Finally, when you think about Python and R, you probably think about pandas: the R-inspired Python library designed with data analysis and modeling in mind. One of the fundamental data structures in pandas is (surprise) the data frame. It's quite easy to convert backward and forward between R and pandas, as follows:

```
import pandas.rpy.common as pd_common
pd_yri_ceu = pd_common.load_data('yri_ceu')
del pd_yri_ceu['PAIRED_FASTQ']
no_paired = pd_common.convert_to_r_dataframe(pd_yri_ceu)
robjects.r.assign('no.paired', no_paired)
robjects.r("print(colnames(no.paired))")
```

   ❑ We start by importing the necessary conversion module. We then convert the R data frame (note that we are converting the `yri_ceu` in the R namespace, not the one on the Python namespace). We delete the column that indicates the name of the paired FASTQ file on the pandas data frame and copy it back to the R namespace. If you print the column names of the new R data frame, you will see that `PAIRED_FASTQ` is missing.

   ❑ As this book enters production, the `pandas.rpy` module is being deprecated (although it's still available).

In the interests of maintaining the momentum of the book, we will not delve into pandas programming (there are plenty of books on this), but I recommend that you take a look at it, not only in the context of interfacing with R, but also as a very good library for data management of complex datasets.

There's more...

It's worth repeating that the advancements on the Python software ecology are occurring at a breakneck pace. This means that if a certain functionality is not available today, it might be released sometime in the near future. So, if you are developing a new project, be sure to check for the very latest developments on the Python front before using a functionality from an R package.

There are plenty of R packages for bioinformatics in the Bioconductor project (http://www.bioconductor.org/). This should probably be your first port of call in the R world for bioinformatics functionalities. However, note that there are many R bioinformatics packages that are not on Bioconductor, so be sure to search the wider R packages on CRAN (refer to the Comprehensive R Archive Network at http://cran.r-project.org/).

There are plenty of plotting libraries for Python. matplotlib is the most common library, but you also have a plethora of other choices. In the context of R, it's worth noting that there is a ggplot2-like implementation for Python based on the Grammar of Graphics description language for charts and this is called—surprise-surprise—ggplot! (http://ggplot.yhathq.com/).

See also

▸ There are plenty of tutorials and books on R; check the R web page (http://www.r-project.org/) for documentation.

▸ For Bioconductor, check the documentation at http://manuals. bioinformatics.ucr.edu/home/R_BioCondManual.

▸ If you work with NGS, you might also want to check *High Throughput Sequence Analysis with Bioconductor* at http://manuals.bioinformatics.ucr.edu/ home/ht-seq.

▸ The rpy library documentation is your Python gateway to R at http://rpy. sourceforge.net/.

▸ The Grammar of Graphics is described in a book aptly named *The Grammar of Graphics*, *Leland Wilkinson*, *Springer*.

▸ In terms of data structures, similar functionality to R can be found on the pandas library. You can find some tutorials at http://pandas.pydata.org/pandas-docs/dev/tutorials.html. The book, *Python for Data Analysis*, *Wes McKinney*, *O'Reilly Media*, is also an alternative to consider.

Performing R magic with IPython

You have probably heard of, and maybe used, the IPython Notebook. If not, then I strongly recommend you try it as it's becoming the standard for reproducible science. Among many other features, IPython provides a framework of extensible commands called magics, which allows you to extend the language in many useful ways.

There are magic functions to deal with R. As you will see in our example, it makes R interfacing much more declarative and easy. This recipe will not introduce any new R functionalities, but hopefully, it will make clear how IPython can be an important productivity boost for scientific computing in this regard.

Getting ready

You will need to follow the previous getting ready steps of the rpy2 recipe. You will also need IPython. You can use the standard command line or any of the IPython consoles, but the recommended environment is the notebook.

If you are using our notebooks, open the 00_Intro/R_magic.ipynb notebook. A notebook is more complete than the recipe presented here with more chart examples. For brevity here, we concentrate only on the fundamental constructs to interact with R using magics.

How to do it...

This recipe is an aggressive simplification of the previous one because it illustrates the conciseness and elegance of R magics:

1. The first thing you need to do is load R magics and ggplot2:

```
import rpy2.robjects.lib.ggplot2 as ggplot2
%load_ext rpy2.ipython
```

   - Note that the % starts an IPython-specific directive.
   - Just as a simple example, you can write on your IPython prompt:

   ```
   %R print(c(1, 2))
   ```

   - See how easy it's to execute the R code without using the robjects package. Actually, rpy2 is being used to look under the hood, but it has been made transparent.

2. Let's read the sequence.index file that was downloaded in the previous recipe:

```
%%R
seq.data <- read.delim('sequence.index', header=TRUE,
stringsAsFactors=FALSE)
seq.data$READ_COUNT <- as.integer(seq.data$READ_COUNT)
seq.data$BASE_COUNT <- as.integer(seq.data$BASE_COUNT)
```

   - Note that you can specify that the whole IPython cell should be interpreted as R code (note the double %%). As you can see, there is no need for a function parameter name translation or (alternatively) explicitly call the robjects.r to execute a code.

3. We can now transfer a variable to the Python namespace (where we could have done Python-based operations):

```
seq_data = %R seq.data
```

4. Let's put this data frame back in the R namespace, as follows:

```
%R -i seq_data
%R print(colnames(seq_data))
```

   - The -i argument informs the magic system that the variable that follows on the Python space is to be copied in the R namespace. The second line just shows that the data frame is indeed available in R. We actually did not do anything with the data frame in the Python namespace, but this serves as an example on how to inject an object back into R.

5. The R magic system also allows you to reduce code as it changes the behavior of the interaction of R with IPython. For example, in the ggplot2 code of the previous recipe, you do not need to use the `.png` and `dev.off` R functions, as the magic system will take care of this for you. When you tell R to print a chart, it will magically appear in your notebook or graphical console. For example, the histogram plotting code from the previous recipe is now simply:

```
%%R
bar <- ggplot(seq_data) +  aes(factor(CENTER_NAME)) +
geom_bar() + theme(axis.text.x = element_text(angle = 90,
hjust = 1))

print(bar)
```

R magics makes interaction with R particularly easy. This is true if you think about how cumbersome multiple language integration tends to be.

The notebook has a few more examples, especially with chart printing, but the core of R-magic interaction is explained before.

See also

▶ For basic instructions on IPython magics, see this notebook at `http://nbviewer.ipython.org/github/ipython/ipython/blob/1.x/examples/notebooks/Cell%20Magics.ipynb`

▶ A list of default extensions is available at `http://ipython.org/ipython-doc/dev/config/extensions/`

▶ A list of third-party magic extensions can be found at `https://github.com/ipython/ipython/wiki/Extensions-Index`

Downloading the example code

You can download the example code files from your account at `http://www.packtpub.com` for all the Packt Publishing books you have purchased. If you purchased this book elsewhere, you can visit `http://www.packtpub.com/support` and register to have the files e-mailed directly to you.

2
Next-generation Sequencing

In this chapter, we will cover the following recipes:

- Accessing GenBank and moving around NCBI databases
- Performing basic sequence analysis
- Working with modern sequence formats
- Working with alignment data
- Analyzing data in variant call format (VCF)
- Studying genome accessibility and filtering SNP data

Introduction

Next-generation Sequencing (**NGS**) is one of the fundamental technological developments of the decade in life sciences. **Whole Genome Sequencing** (**WGS**), RAD-Seq, RNA-Seq, Chip-Seq, and several other technologies are routinely used to investigate important biological problems. These are also called high-throughput sequencing technologies with good reason: they generate vast amounts of data that need to be processed. NGS is the main reason for computational biology becoming a "big data" discipline. More than anything else, this is a field that requires strong bioinformatics techniques. There is a very strong demand for professionals with these skillsets.

Here, we will not discuss each individual NGS technique per se (this will require a whole book on its own). We will use an existing WGS dataset and the human 1000 genomes project to illustrate the most common steps necessary to analyze genomic data. The recipes presented here will be easily applicable for other genomic sequencing approaches. Some of them can also be used for transcriptomic analysis (for example, RNA-Seq). The recipes are also species-independent. So, you will be able to apply them to any other species for which you have sequenced data. The biggest difference in processing data from different species is related to genome size, diversity, and the quality of the assembled genome (if it exists for your species). These will not affect the automated Python part of NGS processing much. In any case, we will discuss different genomes in the next chapter.

As this is not an introductory book, you are expected to know at least what FASTA, FASTQ, BAM, and VCF files are. I will also make use of the basic genomic terminology without introducing it (such as exomes, nonsynonymous mutations, and so on). You are required to be familiar with basic Python. We will leverage this knowledge to introduce the fundamental libraries in Python to perform the NGS analysis. Here, we will follow the flow of a standard bioinformatics pipeline.

However, before we delve into real data from a real project, let's get comfortable with accessing existing genomic databases and basic sequence processing. A simple start before the storm.

Accessing GenBank and moving around NCBI databases

Although you may have your own data to analyze, you will probably need existing genomic datasets. Here, we will see how to access these databases at the **National Center for Biotechnology Information** (**NCBI**). We will not only discuss GenBank, but also other databases at NCBI. Many people refer (wrongly) to the whole set of NCBI databases as GenBank, but NCBI includes the nucleotide database and many others, for example, PubMed. As sequencing analysis is a long subject and this book targets intermediate to advanced users, we will not be very exhaustive with a topic that is, at its core, not very complicated. Nonetheless, it's a good warm-up for more complex recipes at the end of this chapter.

Getting ready

We will use Biopython, which you installed in *Chapter 1, Python and the Surrounding Software Ecology*. Biopython provides an interface to Entrez, the data retrieval system made available by NCBI. This recipe is made available in the `01_NGS/Accessing_Databases.ipynb` notebook.

You will be accessing a live API from NCBI. Note that the performance of the system may vary during the day. Furthermore, you are expected to be a "good citizen" while using it. You will find some recommendations at `http://www.ncbi.nlm.nih.gov/books/NBK25497/#chapter2.Usage_Guidelines_and_Requiremen`. Notably, you are required to specify an e-mail address with your query. You should try to avoid large number of requests (100 or more) during peak times (between 9.00 a.m. and 5.00 p.m. American Eastern Time on weekdays) and do not post more than three queries per second (Biopython will take care of this for you). It's not only good citizenship, but you risk getting blocked if you over use NCBI's servers (a good reason to give a real e-mail address because NCBI may try to contact you).

How to do it...

Now, let's see how we can search and fetch data from NCBI databases:

1. We start by importing the relevant module and configuring the e-mail address:

    ```
    from Bio import Entrez, SeqIO
    Entrez.email = 'put@your.email.here'
    ```

    ❑   We will also import the module to process sequences. Do not forget to put the correct e-mail address.

2. We will now try to find the **Cholroquine Resistance Transporter** (**CRT**) gene in *Plasmodium falciparum* (the parasite that causes the deadliest form of malaria) on the nucleotide database: _~ database._

    ```
    handle = Entrez.esearch(db='nucleotide', term='CRT
    [Gene Name] AND "Plasmodium falciparum"[Organism]')
    rec_list = Entrez.read(handle)
    if rec_list['RetMax'] < rec_list['Count']:
        handle = Entrez.esearch(db='nucleotide', term='CRT
    [Gene Name] AND "Plasmodium falciparum"[Organism]',
                                retmax=rec_list['Count'])
        rec_list = Entrez.read(handle)
    ```

    ❑   We start by searching the nucleotide database for our gene and organism (for the syntax of the search string, check the NCBI website). Then, we read the result that is returned. Note that the standard search will limit the number of record references to 20, so if you have more, you may want to repeat the query with an increased maximum limit. In our case, we will actually override the default limit with retmax.

    ❑   The Entrez system provides quite a few sophisticated ways to retrieve large number of results (for more information, check the Biopython or NCBI Entrez documentation). Although you now have the IDs of all records, you still need to retrieve the records proper.

3. Let's now try to retrieve all these records. The following query will download all matching nucleotide sequences from GenBank, which are 281 at the time of writing this book. You probably do not want to do this all the time:

```
id_list = rec_list['IdList']
hdl = Entrez.efetch(db='nucleotide', id=id_list,
    rettype='gb')
```

  ❏ Well, in this case, go ahead and do it. However, be careful with this technique because you will retrieve a large amount of complete records and some of them will have fairly large sequences inside. You risk downloading a lot of data (both are a strain on your side and on NCBI servers).

  ❏ There are several ways around this. One way is to make a more restrictive query and/or download just a few at a time and stop when you have found the one that is enough. The precise strategy will depend on what you are trying to achieve. In any case, we will retrieve a list of records in the GenBank format (which includes sequences plus a lot of interesting metadata).

4. Let's read and parse the result:

```
recs = list(SeqIO.parse(hdl, 'gb'))
```

  ❏ Note that we have converted an iterator (the result of `SeqIO.parse`) to a list. The advantage is that we can use the result as many times as we want (for example, iterate many times over), without repeating the query on the server. This saves time, bandwidth, and server usage if you plan to iterate many times over. The disadvantage is that it will allocate memory for all records. This will not work for very large datasets; you might not want to do this genome-wide like in the next chapter. We will return to this topic in the last part of the book. If you are doing interactive computing, you will probably prefer to have a list (so that you can analyze and experiment with it multiple times), but if you are developing a library, an iterator will probably be the best approach.

5. We will now just concentrate on a single record. This will only work if you used the exact same preceding query:

```
for rec in recs:
    if rec.name == 'KM288867':
        break
```

  ❏ The `rec` variable now has our record of interest. The `rec.description` will contain its human-readable description.

6. Let's now extract some sequence features, which contain information such as gene products and exon positions on the sequence:

```
for feature in rec.features:
    if feature.type == 'gene':
        print(feature.qualifiers['gene'])
```

```
elif feature.type == 'exon':
    loc = feature.location
    print(loc.start, loc.end, loc.strand)
else:
    print('not processed:\n%s' % feature)
```

❑ If the feature type is gene, we will print its name, which will be in the qualifiers dictionary.

❑ We will also print all locations of exons. Exons, as with all features, have locations in this sequence: a start, an end, and the strand from where they are read. While all the start and end positions for our exons are ExactPosition, note that Biopython supports many other types of positions. One type of position is BeforePosition, which specifies that a location point is before a certain sequence position. Another type of position is BetweenPosition, which gives the interval for a certain location start/end. There are quite a few more position types; these are just some examples. Coordinates will be specified in way that you will be able to retrieve the sequence from a Python array with ranges easily, so generally the start will be one before the value on the record and the end will be equal. The issue of coordinate systems will be revisited in future recipes.

❑ For other feature types, we simply print them. Note that Biopython will provide a human-readable version of the feature when you print it.

7. We will now look at the annotations on the record, which is mostly metadata that is not related to the sequence position:

```
for name, value in rec.annotations.items():
    print('%s=%s' % (name, value))
```

❑ Note that some values are not strings; they can be numbers or even lists (for example, the taxonomy annotation is a list).

8. Last but not least, you can access the fundamental piece of information, the sequence:

```
sequence = rec.seq
```

The sequence object will be the main topic of our next recipe.

There's more...

There are many more databases at NCBI. You will probably want to check the **short read archive (SRA)** database if you are working with next-generation sequencing data. The SNP database will contain information on **Single-nucleotide Polymorphisms (SNPs)**, whereas the protein database will have protein sequences, and so on. A full list of databases in Entrez is linked in the *See also* section of this recipe.

Another database that you probably already know about NCBI is PubMed, which includes a list of scientific and medical citations, abstracts, even full texts. You can also access it via Biopython. Furthermore, GenBank records often contain links to PubMed. For example, we can perform this on our previous record, as shown here:

```
from Bio import Medline
refs = rec.annotations['references']
for ref in refs:
    if ref.pubmed_id != '':
        print(ref.pubmed_id)
        handle = Entrez.efetch(db='pubmed', id=[ref.pubmed_id],
                                    rettype='medline', retmode='text')
        records = Medline.parse(handle)
        for med_rec in records:
            print(med_rec)
```

This will take all reference annotations; check whether they have a PubMed identifier and then access the PubMed database to retrieve the records, parse them, and then print them. The output per record is a Python dictionary. Note that there are many references to external databases on a typical GenBank record.

Of course, there are many other biological databases outside NCBI, such as Ensembl (http://www.ensembl.org) and UCSC Genome Bioinformatics (http://genome.ucsc.edu/). The support for many of these databases in Python will vary a lot.

An introductory recipe on biological databases will not be complete without at least a passing reference to BLAST. BLAST is an algorithm that assesses the similarity of sequences. NCBI provides a service that allows you to compare your sequence of interest against its own database. Of course, you can have your local BLAST database instead of using NCBI's service. Biopython provides extensive support for this, but as this is too introductory, I will just refer you to the Biopython tutorial.

See also

▸ You can find more examples on the Biopython tutorial at http://biopython.org/DIST/docs/tutorial/Tutorial.html

▸ A list of accessible NCBI databases can be found at http://www.ncbi.nlm.nih.gov/gquery/

▸ A great Q&A site where you can find help for your problems with databases and sequence analysis is biostars (http://www.biostars.org); you can use it for all the content in this book, not just for this recipe.

Performing basic sequence analysis

We will now do some basic analysis on DNA sequences. We will work with FASTA files and do some manipulation, such as reverse complementing or transcription. As with the previous recipe, we will use Biopython, which you installed in *Chapter 1, Python and the Surrounding Software Ecology*. These two recipes provide you with the necessary introductory building blocks in which we will perform all the modern next-generation sequencing analysis, and then genome processing in this and the next chapter.

Getting ready

If you are using notebooks, then open `01_NGS/Basic_Sequence_Processing.ipynb`. If not, you will need to download a FASTA sequence. We will use the human lactase gene as an example; you can get this using the knowledge you got from the previous recipe using Entrez research interface:

```
from Bio import Entrez, SeqIO
Entrez.email = "your@email.here"
hdl = Entrez.efetch(db='nucleotide', id=['NM_002299'],
    rettype='fasta')  # Lactase gene
seq = SeqIO.read(hdl, 'fasta')
```

Note that our example sequence is available on the Biopython sequence record.

How to do it...

Let's take a look at the following steps:

1. As our sequence of interest is available in a Biopython sequence object, let's start by saving it to a FASTA file on our local disk:

    ```
    from Bio import SeqIO
    w_hdl = open('example.fasta', 'w')
    w_seq = seq[11:5795]
    SeqIO.write([w_seq], w_hdl, 'fasta')
    w_hdl.close()
    ```

Note that we chose a subset of a sequence to write what was hardcoded in our known case. If you look at the features of the downloaded sequence, you will see that it corresponds to the **Coding Sequence** (**CDS**) part. When you download a sequence, it may have more than the code part of the gene. It probably has the coding sequence and exons. This includes everything that is not removed by RNA splicing (which is more than the coding sequence) and many other features. In this case, we have chosen a sequence with only a single CDS entry, but in general, you may have to go through multiple CDS features in order to reconstruct the coding sequence (that is, start and end codons plus all codons coding amino acids). So, do not forget that the downloaded "gene sequence" is normally bigger than the exomic part, which is bigger than the coding (CDS) part.

The `SeqIO.write` function takes a list of sequences to write (not just a single one). Be careful with this idiom. If you want to write many sequences (easily millions with NGS), do not use a list, as shown in the preceding code because this will allocate massive amounts of memory. Either use an iterator or use the `SeqIO.write` function several times with a subset of sequence on each write.

2. In most situations, you will actually have the sequence on the disk, so you will be interested in reading it:

```
recs = SeqIO.parse('example.fasta', 'fasta')
for rec in recs:
    seq = rec.seq
    print(rec.description)
    print(seq[:10])
    print(seq.alphabet)
```

- Here, we are concerned with processing a single sequence, but FASTA files can contain multiple records. The Python idiom to perform this is quite easy. To read a FASTA file, you just use standard iteration techniques.

- For our example, the preceding code will print:

```
gi|32481205|ref|NM_002299.2| Homo sapiens lactase (LCT),
mRNA

GTTCCTAGAA

SingleLetterAlphabet()
```

- ❑ The first line is the FASTA textual descriptor (remember that FASTA files contain a description line followed by a sequence for an arbitrary number of sequences). Although, in theory, this can be anything, it's common that it has some structure. For example, in the preceding NCBI example, you have the unique `gi` number, followed by the reference sequence ID and a textual description. This means that in theory, the arbitrary description string is most probably amenable to automated parsing. If you do not know how a FASTA file looks, just open the `example.fasta` file with a text editor.

- ❑ Note that we printed `seq[:10]`. The sequence object can use typical array slices to get part of a sequence.

3. We will now change the alphabet of our sequence:

```
from Bio import Seq
from Bio.Alphabet import IUPAC
seq = Seq.Seq(str(seq), IUPAC.unambiguous_dna)
```

- ❑ Probably the biggest value of the sequence object (compared to a simple string) comes from the alphabet information. The sequence object will be able to impose useful constraints and operations on the underlying string, based on the expected alphabet. The original alphabet on the FASTA file is not very informative, but in this case, we know that we have a DNA alphabet. So, we create a new sequence with a more informative alphabet.

4. As we now have an unambiguous DNA, we can transcribe it as follows:

```
rna = Seq.Seq(str(seq), IUPAC.unambiguous_dna)
rna = seq.transcribe()
print(rna)
```

> Note that the `Seq` constructor takes a string, not a sequence. You will see that the alphabet of the `rna` variable is now `IUPACUnambigousRNA`.

5. Finally, we can translate our gene into a protein:

```
prot = seq.translate()
print(prot)
```

Now, we have a protein alphabet with the annotation that there is a stop codon (so, our protein is complete).

There's more...

Much more can be said about the management of sequence in Biopython, but this is mostly introductory material that you can find in the Biopython tutorial. I think it's important to give you a taste of sequence management, mostly for completion purposes. In order to support readers who might have some experience in other bioinformatics' fields, but are actually starting with sequence analysis, there are nonetheless a few points that you should be aware of:

▸ When you perform an RNA translation to get your protein, be sure to use the correct genetic code. Even if you are working with "common" organisms (such as humans), remember that the mitochondrial genetic code is different.

▸ Biopython's `Seq` object is much more flexible than is shown here. For some good examples, refer to the Biopython tutorial. However, this recipe will be enough for the work we need to do with FASTQ files (see the next recipe).

▸ To deal with strand-related issues there are, as expected, sequence functions like `reverse_complement`.

See also

▸ Genetic codes known to Biopython are the ones specified by NCBI at `http://www.ncbi.nlm.nih.gov/Taxonomy/Utils/wprintgc.cgi`

▸ As in the previous recipe, the Biopython tutorial is your main port of call and is available at `http://biopython.org/DIST/docs/tutorial/Tutorial.html`

▸ Be sure to also check the Biopython SeqIO page at `http://biopython.org/wiki/SeqIO`

Working with modern sequence formats

Here, we will work with FASTQ files, the standard format output by modern sequencers. You will learn how to work with quality scores per base and also consider the variations in output coming from different sequencing machines and databases. This is the first recipe that will use real data (big data) from the human 1000 genomes project. We will start with a brief description of the project.

Getting ready

The human 1000 genomes project aims to catalog world-wide human genetic variation and takes advantage of modern sequencing technology to do WGS. This project makes all data publicly available, which includes output from sequencers, sequence alignments, and SNP calls, among many other artifacts. The name 1000 genomes is actually a misnomer because it currently includes more than 2500 samples. These samples are divided into 26 populations, spanning the whole planet. We will mostly use data from four populations: African Yorubans (YRI), Utah Residents with Northern and Western European Ancestry (CEU), Japanese in Tokyo (JPT), and Han Chinese in Beijing (CHB). The reason we chose these specific populations is because they were the first ones that came from HapMap, an old project with similar goals. They used genotyping arrays to know more about the quality of this subset. We will revisit the 1000 genomes and HapMap projects in the chapter on population genetics.

Next-generation datasets are generally very large. As we will be using real data, some of the files that you will download will be big. While I have tried to choose the smallest real examples possible, you will still need a good network connection and considerably large amount of disk space. Waiting for the download will probably be your biggest hurdle in this case, but data management is a serious problem with NGS. In real life, you will need to budget time for data transfer, allocate disk space (which can be financially costly), and consider backup policies. The most common initial mistake with NGS is to think that these problems are trivial, but they are not. An operation like copying a set of BAM files to a network, or even on your computer, will become a headache. Be prepared. After downloading large files, at the very least, you should check that the size is correct. Some databases offer MD5 checksums. You can compare these checksums with the ones on the files you downloaded by using tools like md5sum.

If you use notebooks, do not forget to download the data, as specified on the first cell of `01_NGS/Working_with_FASTQ.ipynb`. If not, download the file `SRR003265.filt.fastq.gz`, which is linked in `https://github.com/tiagoantao/bioinf-python/blob/master/notebooks/Datasets.ipynb`. This is a fairly small file (27 MB) and represents part of a sequenced data of a Yoruban female (NA18489). If you refer to the 1000 genomes project, you will see that vast majority of FASTQ files are much bigger (up to 2 orders of magnitude bigger).

The processing of FASTQ sequence files will mostly be performed (using Biopython). Let's do it.

How to do it...

Before we start coding, let's take a look at the FASTQ file, in which you will have many records, as shown in the following code:

```
@SRR003258.1 30443AAXX:1:1:1053:1999 length=51
ACCCCCCCCCACCCCCCCCCCCCCCCCCCCCCCCCCCCCCACACACACCAACAC
+
=IIIIIIIII5IIIIIII>IIII+GIIIIIIIIIIIIII(IIIII01&III
```

Line 1 starts with an @, followed by a sequence identifier and a description string. The description string will vary from a sequencer or a database source, but will normally be amenable to automated parsing.

The second line has the sequence DNA, which is just like a FASTA file. The third line is a +, sometimes followed by the description line on the first line.

The fourth line contains quality values for each base read on line two. Each letter encodes a Phred quality score (http://en.wikipedia.org/wiki/Phred_quality_score), which assigns a probability of error to each read. This encoding can vary a bit among platforms. Be sure to check for this on your specific platform.

1. Let's now open the file:

    ```
    import gzip
    from Bio import SeqIO
    recs = SeqIO.parse(gzip.open('SRR003265.filt.fastq.gz'),
    'fastq')
    rec = next(recs)
    print(rec.id, rec.description, rec.seq)
    print(rec.letter_annotations)
    ```

    ❏   We will open a gzip file so that we can use the Python gzip module. We will also specify the fastq format. Note that some variations in this format will impact the interpretation of the Phred quality scores. You may want to specify a slightly different format. Refer to http://biopython.org/wiki/SeqIO for all formats.

> You should usually store your FASTQ files in a compressed format. Not only do you gain a lot of disk space, as these are text files, but you probably also gain some processing time. Although decompressing is a slow process, it can still be faster than reading a much bigger (uncompressed) file from a disk.

    ❏   We print standard fields and quality scores from the previous recipe in rec.letter_annotations. As long as we choose the correct parser, Biopython will convert all the Phred encoding letters to logarithmic scores, to be used soon.

❏ Now, do not do this:

```
recs = list(recs)    # do not do it!
```

❏ Although, this might work with some FASTA files (and with this very small FASTQ file), if you do something like this, you will allocate memory to load the complete file in memory. With an average FASTQ file, this is the best way to crash your computer. As a rule, always iterate over your file. If you have to perform several operations over it, you have two main options. The first option is perform a single iteration with all operations at once. The second option is open a file several times and repeat the iteration.

2. Now, let's take a look at the distribution of nucleotide reads:

```
from collections import defaultdict
recs = SeqIO.parse(gzip.open('SRR003265.filt.fastq.gz'),
    'fastq')
cnt = defaultdict(int)
for rec in recs:
    for letter in rec.seq:
        cnt[letter] += 1
tot = sum(cnt.values())
for letter, cnt_value in cnt.items():
    print('%s: %.2f %d' % (letter, 100. * cnt_value / tot,
    cnt_value))
```

❏ We will reopen the file again and use a `defaultdict` to maintain a count of nucleotide references in the FASTQ file. If you have never used this Python standard dictionary type, you may want to consider it because it removes the need to initialize dictionary entries, assuming default values for each type.

Note that there is a residual number for N calls. These are calls in which a sequencer reports an unknown base. In our FASTQ file example, we have cheated a bit because we used a filtered file (the fraction of Ns will be quite low). Expect a much bigger number of N calls in a file that came out of the sequencer unfiltered. In fact, you can even expect something more with regards to the spatial distribution of N calls.

3. Let's plot the distribution of Ns according to its read position:

```
%matplotlib inline
import seaborn as sns
import matplotlib.pyplot as plt
recs = SeqIO.parse(gzip.open('SRR003265.filt.fastq.gz'),
    'fastq')
n_cnt = defaultdict(int)
for rec in recs:
```

```
    for i, letter in enumerate(rec.seq):
        pos = i + 1
        if letter == 'N':
            n_cnt[pos] += 1
seq_len = max(n_cnt.keys())
positions = range(1, seq_len + 1)
fig, ax = plt.subplots()
ax.plot(positions, [n_cnt[x] for x in positions])
ax.set_xlim(1, seq_len)
```

- ❑ The first line only works on IPython (you should remove it on a standard Python) and it will inline any plots. We then import the `seaborn` library. Although, we do not use it explicitly at this point, this library has the advantage of making matplotlib plots look better because it tweaks the default matplotlib style.

- ❑ We then open the file to parse again (remember that you do not use a list, but iterate again). We iterate through the file and get the position of any references to N. Then, we plot the distribution of Ns as a function of distance to start the sequence:

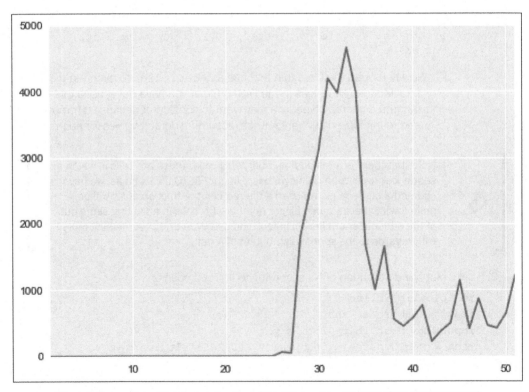

Figure 1: The number of N calls as a function of the distance from the start of the sequencer read

- ❏ You will see that until position 25, there are no errors. This is not what you will get from a typical sequencer output. Our example file is already filtered and the 1000 genomes filtering rules enforce that no N calls can occur before position 25.

- ❏ While we cannot study the behavior of Ns in this dataset before position 25 (feel free to use one of your own unfiltered FASTQ files with this code in order to see how Ns distribute across the read position), we can see that after position 25, the distribution is far from uniform. There is an important lesson here, which is that the quantity of uncalled bases is position dependent. So, what about the quality of reads?

4. Let's study the distribution of Phred scores (that is, the quality of our reads):

```
recs = SeqIO.parse(gzip.open('SRR003265.filt.fastq.gz'),
'fastq')
cnt_qual = defaultdict(int)
for rec in recs:
        for i, qual in enumerate(
        rec.letter_annotations['phred_quality']):
        if i < 25:
            continue
        cnt_qual[qual] += 1
tot = sum(cnt_qual.values())
for qual, cnt in cnt_qual.items():
    print('%d: %.2f %d' % (qual, 100. * cnt / tot, cnt))
```

- ❏ We start by reopening the file (again) and initializing a default dictionary. We then get the `phred_quality` letter annotation, but we ignore sequencing positions that are up to 24 base pairs from the start (because of the filtering of our FASTQ file, if you have an unfiltered file, you may want to drop this rule). We add the quality score to our default dictionary and finally print it.

As a short reminder, the Phred quality score is a logarithmic representation of the probability of an accurate call. This probability is given as $10^{(-Q/10)}$. So, a Q of 10 represents a 90 percent call accuracy, 20 represents 99 percent call accuracy, and 30 will be 99.9 percent. For our file, the maximum accuracy will be 99.99 percent (40). In some cases, values of 60 are possible (99.9999 percent accuracy).

5. More interestingly, we can plot the distribution of qualities according to their read position:

```
recs = SeqIO.parse(gzip.open('SRR003265.filt.fastq.gz'),
    'fastq')
qual_pos = defaultdict(list)
for rec in recs:
    for i, qual in enumerate(rec.letter_annotations['phred_
quality']):
        if i < 25 or qual == 40:
            continue
        pos = i + 1
        qual_pos[pos].append(qual)
vps = []
poses = qual_pos.keys()
poses.sort()
for pos in poses:
    vps.append(qual_pos[pos])
fig, ax = plt.subplots()
sns.boxplot(vps, ax=ax)
ax.set_xticklabels([str(x) for x in range(26,
    max(qual_pos.keys()) + 1)])
```

- ❑ In this case, we will ignore both positions sequenced 25 base pairs from start (again, remove this rule if you have unfiltered sequencer data) and the maximum quality score for this file (40). However, in your case, you can consider starting your plotting analysis also with the maximum. You may want to check the maximum possible value for your sequencer hardware. Generally, as most calls can be performed with maximum quality, you may want to remove them if you are trying to understand where quality problems lie.

- ❑ Note that we are using seaborn's `boxplot` function; we use this only because the output looks slightly better than the standard matplotlib `boxplot`. If you prefer not to depend on seaborn, just use the stock matplotlib function. In this case, you will call `ax.boxplot(vps)` instead of `sns.boxplot(vps, ax=ax)`.

❑ As expected, the distribution of qualities is not uniform, as shown in the following figure:

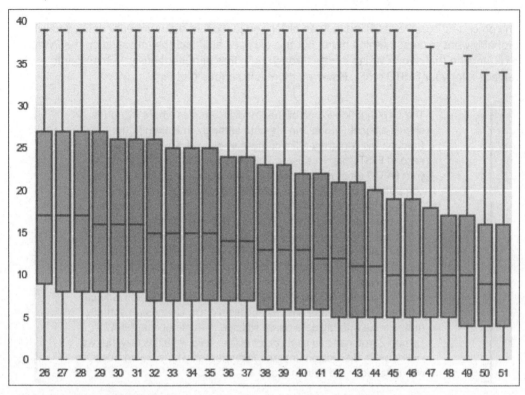

Figure 2: The distribution of Phred scores as a function of the distance from the start of the sequencer read

There's more...

Although it's impossible to discuss all the variations of output coming from sequencer files, paired-end reads are worth mentioning because they are common and require a different processing approach. With paired-end sequencing, both ends of a DNA fragment are sequenced with a gap in the middle (called the insert). In this case, you will have two files produced: X_1.FASTQ and X_2.FASTQ. Both files will have the same order and exact same number of sequences. The first sequence in X_1 pairs with the first sequence of X_2, and so on. With regards to the programming technique, if you want to keep the pairing information, you might be tempted to perform something like this:

```
f1 = gzip.open('X_1.filt.fastq.gz')
f2 = gzip.open('X_2.filt.fastq.gz')
recs1 = SeqIO.parse(f1, 'fastq')
recs2 = SeqIO.parse(f2, 'fastq')
cnt = 0
```

```
for rec1, rec2 in zip(recs1, recs2):
    cnt +=1
print('Number of pairs: %d' % cnt)
```

The preceding code reads all pairs in an order and just count the number of pairs. You will probably want to do something more, but this exposes a dialect that is based on the Python `zip` function that allows you to iterate through both files simultaneously. Remember to replace X for your FASTQ prefix. However, there is a serious problem.

The `zip` function is one of the (many) examples where Python 3 shines against Python 2 in big data settings. In Python 2, the `zip` function will construct a list composed of pairs for the first and the second FASTQ file. This list will be constructed in-memory with all your FASTQ records from both files and will most probably crash your computer. You should avoid doing this in Python2. This is when Python 3 comes into picture. In this case, `zip` has different semantics: it will only return an iterator (not a list) and elements as you consume them. In Python 3, the preceding code will perform handsomely. Python 3 semantics are typically lazier than Python 2, which will only return values when they are needed. It will not construct gruesome lists in-memory. If you want the list of enumerators, by the way, this is quite easy to do in Python 3; just cast: `list(zip(...))`. Now, you can have a permanent and directly accessible list. Laziness is actually quite useful with big data as we will see in the final chapter. With Python 2, you can still use `itertools.zip`, which is lazy, but we will defer its discussion until the last chapter, where we address some contrasts between Python 2 and 3.

With Python 2 and without `itertools`, you can use a less elegant dialect. Replace `for`, as shown in the preceding code, with this:

```
for rec1 in recs1:
    next(recs2)
    cnt +=1
```

Note that you might not care much for the pair order. In this case, you just open both files sequentially with a simpler processing strategy.

Finally, if you are sequencing human genomes, you may want to use sequencing data from Complete Genomics. In this case, read the *There's more...* section in the next recipe, where we briefly discuss Complete Genomics data.

See also

> ▶ The Wikipedia page (`http://en.wikipedia.org/wiki/FASTQ_format`) on the FASTQ format is quite informative

- ▶ You can find more information on the 1000 genomes project at `http://www.1000genomes.org/`

- ▶ Information about the Phred quality score can be found at `http://en.wikipedia.org/wiki/Phred_quality_score`

- ▶ Illumina provides a good introduction page to paired-end reads at `http://technology.illumina.com/technology/next-generation-sequencing/paired-end-sequencing_assay.html`

- ▶ The paper *Computational methods for discovering structural variation with next-generation sequencing* from *Medvedev et al* on *Nature Methods* (`http://www.nature.com/nmeth/journal/v6/n11s/abs/nmeth.1374.html`); note that this is not open access.

Working with alignment data

After you receive your data from the sequencer, you will normally use a tool such as bwa to align your sequences to a reference genome. Most users will have a reference genome for their species. You can read more on reference genomes in the next chapter.

The most common representation for aligned data is the **Sequence Alignment/Map** (**SAM**) format. Due to the massive size of most of these files, you will probably work with its compressed version (BAM). The compressed format is indexable for extremely fast random access (for example, to speedily find alignments to a certain part of a chromosome). Note that you will need to have an index for your BAM file normally created by the tabix utility of samtools. Samtools is probably the most widely used tool to manipulate SAM/BAM files.

Getting ready

As discussed in the previous recipe, we will use data from the 1000 genomes project. We will use the exome alignment for chromosome 20 of female NA18489. This is "just" 312 MB. The whole exome alignment for this individual is 14.2 GB and the whole genome alignment (at a low coverage of 4x) is 40.1 GB. This data is paired-end with reads of 76 bp, which is common nowadays, but slightly more complex to process. We will take this into account. If your data is not paired, just simplify the following recipe appropriately.

As usual, if you use notebooks, the cell at the top of `01_NGS/Working_with_BAM.ipynb` will download the data for you. If you don't use notebooks, get it from our dataset list at `https://github.com/tiagoantao/bioinf-python/blob/master/notebooks/Datasets.ipynb` the files `NA18490_20_exome.bam` and `NA18490_20_exome.bam.bai`.

We will use Pysam, a Python wrapper to the samtools C API. This was installed in *Chapter 1, Python and the Surrounding Software Ecology*.

How to do it...

Before you start coding, note that you can inspect the BAM file using `samtools view -h` (if you have samtools installed, which we recommend, even if you use GATK or something else for variant calling). We suggest that you take a look at the header file and the first few records. The SAM format is too complex to be described here. There is plenty of information on the Internet about it; nonetheless, sometimes, there are really interesting information buried in these header files.

One of the most complex operations in NGS is to generate good alignment files from a raw sequence data. It not only calls the aligner, but also cleans up data. Now, in the @PG headers of high quality BAM files, you will find the actual command lines used for most, if not all, of the procedure used to generate this BAM file. In our example BAM file, you will find all the information needed to run bwa, samtools, the GATK indel realigner, and the Picard application suite to clean up data. Remember that while you can generate BAM files easily, the programs after it will be quite picky in terms of correctness of the BAM input. For instance, if you use GATK's variant caller to generate genotype calls, the files will have to be extensively cleaned. The header of other BAM files can thus provide you with the best way to generate yours. A final recommendation is that if you do not work with human data, try to find good BAMs for your species because the parameters of a program may be slightly different. Also, if you use other than the WGS data, check for similar types of sequencing data.

Take a look at the following steps:

1. Let's inspect the header files:

```
import pysam
bam = pysam.AlignmentFile('NA18489.chrom20.ILLUMINA.bwa.YRI.exome
.20121211.bam', 'rb')
headers = bam.header
for record_type, records in headers.items():
    print (record_type)
    for i, record in enumerate(records):
        print('\t%d' % (i + 1))
        if type(record) == str:
            print('\t\t%s' % record)
        elif type(record) == dict:
            for field, value in record.items():
                print('\t\t%s\t%s' % (field, value))
```

- ❑ The header is represented as a dictionary (where the key is the record type). As there can be several instances of the same record type, the value of the dictionary is a list (where each element is again a dictionary or sometimes a string containing tag/value pairs).

2. We will now inspect a single record. The amount of data per record is quite complex. Here, we will focus on some of the fundamental fields for paired-end reads. Check the SAM file specification and the Pysam API documentation for more details:

```
for rec in bam:
    if rec.cigarstring.find('M') > -1 and \
        rec.cigarstring.find('S') > -1 and \
        not rec.is_unmapped and \
        not rec.mate_is_unmapped:
        break
print(rec.query_name, rec.reference_id,
bam.getrname(rec.reference_id), rec.reference_start,
    rec.reference_end)
print(rec.cigarstring)
print(rec.query_alignment_start, rec.query_alignment_end,
    rec.query_alignment_length)
print(rec.next_reference_id, rec.next_reference_start,
    rec.template_length)
print(rec.is_paired, rec.is_proper_pair, rec.is_unmapped,
    rec.mapping_quality)
print(rec.query_qualities)
print(rec.query_alignment_qualities)
print(rec.query_sequence)
```

- Note that the BAM file object is iterable over its records. We will transverse it until we find a record whose CIGAR string contains a match and a soft clip.

- The CIGAR string gives an indication of the alignment of individual bases. The clipped part of the sequence is a part that the aligner failed to align (but is not removed from the sequence). We will also want the read, its mate ID, and position (the pair, as we have paired-end reads) that was mapped to the reference genome.

- First, we print the query template name followed by the reference ID. The reference ID is a pointer to name of the sequence on the given references on the lookup table of references. An example will make this clear. For all records on this BAM file, the reference ID is 19 (a noninformative number), but if you apply `bam.getrname(19)`, you will get 20, which is the name of the chromosome. So, do not confuse the reference ID (in this case, 19) with the name of the chromosome (20). This is then followed by the reference start and reference end. Pysam is 0-based, not 1-based. So, be careful when you convert coordinates to other libraries. You will notice that the start and end for this case is 59996 and 60048, which means an alignment of 52 bases. Why only 52 bases when the read size is 76 (remember, the read size used in this BAM file). The answer can be found on the CIGAR string, which in our case will be 52M24S, which is a 52 bases match, followed by 24 bases that were soft-clipped.

❏ Then, we print where the alignment starts and ends and calculate its length. By the way, you can compute this by looking at the CIGAR string. It starts at 0 (as the first part of the read is mapped) and ends at 52. The length is 76 again.

❏ Now, we query the mate (something that you will only do if you have paired-end reads). We get its reference ID (as shown in the previous code), its start position, and a measure of the distance between both pairs. This measure of distance only makes sense if both mates are mapped to the same chromosome.

❏ We then plot the Phred score (refer to the previous recipe on Phred scores) for the sequence, and then only for the aligned part.

❏ Finally, we print the sequence (not to forget this!). This is the complete sequence, not the clipped one (of course, you can use the preceding coordinates to clip).

3. Let's now plot the distribution of successfully mapped positions in a subset of sequences in the BAM file:

```
%matplotlib inline
import seaborn as sns
import matplotlib.pyplot as plt
counts = [0] * 76
for n, rec in enumerate(bam.fetch('20', 0, 10000000)):
    for i in range(rec.query_alignment_start, rec.query_alignment_
end):
        counts[i] += 1
freqs = [x / (n + 1.) for x in counts]
plt.plot(range(1, 77), freqs)
```

❏ We perform the usual graphical import (remove the first line if you are not on IPython). We start by initializing an array to keep the count for the entire 76 positions. Note that we then fetch only the records for chromosome 20 between positions 0 and 10 Mbp. We will just use a small part of the chromosome here. It's fundamental to have an index (generated by tabix) for these kinds of fetch operations; the speed of execution will be completely different.

❏ We traverse all records in the 10 Mbp boundary. For each boundary, we get the alignment start and end and increase a counter of mappability among positions that were aligned. We finally convert this to frequencies and then plot it, as shown in the following figure:

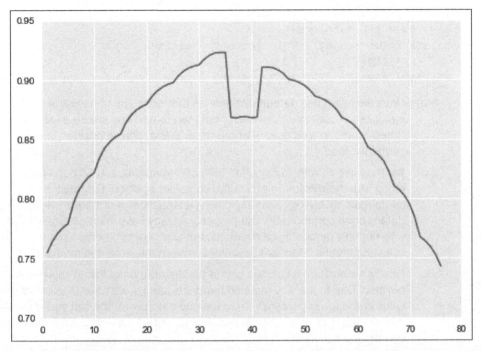

Figure 3: The percentage of mapped calls as a function of the position from the start of the sequencer read

- ❑ It's quite clear that the distribution of mappability is far from being uniform; it's worse at the extremes, with a drop in the middle.

4. Finally, let's get the distribution of Phred scores across the mapped part of the reads. As you may suspect, this is probably not going to be uniform:

```
from collections import defaultdict
import numpy as np
phreds = defaultdict(list)
for rec in bam.fetch('20', 0, None):
    for i in range(rec.query_alignment_start,
    rec.query_alignment_end):
        phreds[i].append(rec.query_qualities[i])

maxs = [max(phreds[i]) for i in range(76)]
tops = [np.percentile(phreds[i], 95) for i in range(76)]
medians = [np.percentile(phreds[i], 50) for i in range(76)]
bottoms = [np.percentile(phreds[i], 5) for i in range(76)]

medians_fig = [x - y for x, y in zip(medians, bottoms)]
tops_fig = [x - y for x, y in zip(tops, medians)]
```

```
maxs_fig = [x - y for x, y in zip(maxs, tops)]
fig, ax = plt.subplots()
ax.stackplot(range(1, 77), (bottoms, medians_fig,
    tops_fig))
ax.plot(range(1, 77), maxs, 'k-')
```

- ❏ Here, we again use default dictionaries that allow you to have a few initialization code. We now fetch from start to end and create a list of Phred scores in a dictionary whose index is the relative position in the sequence read.

- ❏ We then use NumPy to calculate 95th, 50th (median), and 5th percentiles along with the maximum of quality scores per position. For most computational biology analysis, having a statistical summarized view of the data is quite common. So, you probably already have yourself familiarized with not only percentile calculations, but also other Pythonic ways to calculate means, standard deviations, maximums, and minimums.

- ❏ Finally, we perform a stacked plot of the distribution of Phred scores per position. Due to the way matplotlib expects stacks, we have to subtract the value of the lower percentile from the one before with the call stackplot. We can use the list for the bottom percentiles, but we have to correct the median and the top as follows:

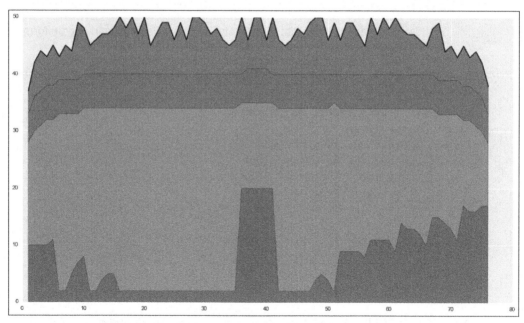

Figure 4: The distribution of Phred scores as a function of the position in the read; the bottom blue color spans from 0 to the 5th percentile; the green color up to the median, red to the 95th percentile, and purple to the maximum

There's more...

Although we will discuss data filtering in another recipe in this chapter, it's not our objective to explain the SAM format in detail or give a detailed course in data filtering. This task will require a book of its own, but with the basics of Pysam, you can navigate through SAM/BAM files. However, in the last recipe, we will take a look at extracting genome-wide metrics from BAM files (via annotations on VCF files that represent metrics of BAM files) for the purpose of understanding the overall quality of our dataset.

You will probably have very large data files to work with. It's possible that some BAM processing will take too much time. One of the first approaches to reduce the computation time is subsampling. For example, if you subsample at 10 percent, you ignore nine records out of 10. For many tasks, such as some of the analysis done for quality assessment of BAM files, subsampling at 10 percent (or even 1 percent) will be enough to have a grasp of the quality of the file.

If you use human data, you may have your data sequenced at Complete Genomics. In this case, the alignment files will be different. Although Complete Genomics provides tools to convert to standard formats, you might be served better using their own data.

See also

- The SAM/BAM format is described at `http://samtools.github.io/hts-specs/SAMv1.pdf`

- You can find an introductory explanation to the SAM format on the Abecassis' group wiki page at `http://genome.sph.umich.edu/wiki/SAM`

- If you really need to get complex statistics from BAM files, Alistair Miles' pysamstats library is your port of call at `https://github.com/alimanfoo/pysamstats`

- To convert your raw sequence data to alignment, data you will need an aligner, the most widely used is the Burrows-Wheeler Aligner, BWA (`http://bio-bwa.sourceforge.net/`)

- Picard (surely a Star Trek – the next-generation reference) is the most commonly used tool to clean up BAM files; refer to `http://broadinstitute.github.io/picard/`

- The technical forum for sequence analysis is SEQanswers (`http://seqanswers.com/`)

- I would like to repeat the recommendation on biostars here (referred in the previous recipe); it's a treasure trove of information and has a very friendly community at `http://www.biostars.org/`

- If you have the complete genomics data, take a look at their data results FAQ at `http://www.completegenomics.com/FAQs/Data-Results/`

Analyzing data in the variant call format

After running a genotype caller (for example, GATK or samtools), you will have a **variant call format** (**VCF**) file reporting on genomic variations, such as single-nucleotide polymorphisms (**SNPs**), **Insertions/Deletions** (**INDELs**), **copy number variation** (**CNVs**), and so on. In this recipe, we will discuss VCF processing with the PyVCF module.

Getting ready

While next-generation sequencing is all about big data, there is a limit to how much I can ask you to download as a dataset for this book. I believe that 2 to 20 GB of data for a tutorial is asking too much. While the 1000 genomes' VCF files with realistic annotations are in this order of magnitude, we will want to work with much less data here. Fortunately, the bioinformatics community has developed tools to allow partial download of data. As part of the samtools/htslib package (http://www.htslib.org/), you can download tabix and bgzip, which will take care of data management (on Debian, Ubuntu, and Linux, just go to sudo apt-get install tabix to install). On the command line, perform the following:

```
tabix -fh ftp://ftp-
trace.ncbi.nih.gov/1000genomes/ftp/release/20130502/supporting/vcf_
with_sample_level_annotation/ALL.chr22.phase3_shapeit2_mvncall_i
ntegrated_v5_extra_anno.20130502.genotypes.vcf.gz 22:1-17000000
|bgzip -c > genotypes.vcf.gz

tabix -p vcf genotypes.vcf.gz
```

If the preceding link does not work, be sure to check the dataset page at https://github.com/tiagoantao/bioinf-python/blob/master/notebooks/Datasets.ipynb for an update.

The first line will partially download the VCF file for chromosome 22 (up to 17 Mbp) of the 100 genomes project. Then, bgzip will compress it.

The second line will create an index, which we will need for direct access to a section of the genome.

As usual, you have the code to do this on a notebook 01_NGS/Working_with_VCF.ipynb.

How to do it...

Take a look at the following steps:

1. Let's start by inspecting the information that we can get per record:

```
import vcf
v = vcf.Reader(filename='genotypes.vcf.gz')

print('Variant Level information')
infos = v.infos
for info in infos:
    print(info)

print('Sample Level information')
fmts = v.formats
for fmt in fmts:
    print(fmt)
```

- We start by inspecting annotations available for each record (remember that each record encodes a variant, such as SNP, CNV, Indel, and so on, and the state of that variant per sample). At the variant (record) level, we find AC: the total number of ALT alleles in called genotypes, AF: the estimated allele frequency, NS: the number of samples with data, AN: the total number of alleles in called genotypes, and DP: the total read depth among others. There are others, but they are mostly specific to the 1000 genomes project (here, we will try to be as general as possible). Your own dataset may have more annotations (or none of these).

- At the sample level, there are only two annotations in this file: GT: genotype and DP: the per sample read depth. You have the per variant (total) read depth and per sample read depth; be sure not to confuse both.

2. Now that we know which information is available, let's inspect a single VCF record:

```
v = vcf.Reader(filename='genotypes.vcf.gz')
rec = next(v)
print(rec.CHROM, rec.POS, rec.ID, rec.REF, rec.ALT,
    rec.QUAL, rec.FILTER)
print(rec.INFO)
print(rec.FORMAT)
samples = rec.samples
print(len(samples))
sample = samples[0]
print(sample.called, sample.gt_alleles, sample.is_het,
    sample.phased)
print(int(sample['DP']))
```

- ❑ We start by retrieving the standard information: the chromosome, position, identifier, reference base, (typically just one), alternative bases (can have more than one, but it's not uncommon as a first filtering approach to only accept a single ALT, for example, only accept bi-allelic SNPs), quality (as you might expect, Phred-scaled), and filter status. Regarding the filter status, remember that whatever the VCF file says, you may still want to apply extra filters (as in the next recipe).

- ❑ We then print the additional variant-level information (AC, AS, AF, AN, DP, and so on) followed by the sample format (in this case, DP and GT). Finally, we count the number of samples and inspect a single sample to check whether it was called for this variant. If available, the reported alleles, heterozygosity and phasing status (this dataset happens to be phased, which is not that common).

3. Let's check the type of variant and the number of non-biallelic SNPs in a single pass:

```python
from collections import defaultdict
f = vcf.Reader(filename='genotypes.vcf.gz')

my_type = defaultdict(int)
num_alts = defaultdict(int)

for rec in f:
    my_type[rec.var_type, rec.var_subtype] += 1
    if rec.is_snp:
        num_alts[len(rec.ALT)] += 1
print(num_alts)
print(my_type)
```

- ❑ We use the now common Python default dictionary. We find that this dataset has INDELs (both insertions and deletions), CNVs, and, of course, SNPs (roughly two-thirds being transitions with one-third transversions). There is a residual number (79) of tri-allelic SNPs.

There's more...

The purpose of this recipe is to get you up to speed on the PyVCF module. At this stage, you should be comfortable with the API. We will not spend too much time here on usage details because this will be the main purpose of the next recipe: using the VCF module to study the quality of your variant calls.

It will probably not be a shocking revelation that PyVCF is not the fastest module on earth. The file format (highly text-based) makes processing a time-consuming task. There are two main strategies of dealing with this problem. One strategy is parallel processing, which we will discuss in the last chapter. The second strategy is converting to a more efficient format; we will provide an example for this in *Chapter 4, Population Genetics*. Note that VCF developers are doing a binary (BCF) version to deal with part of these problems (`http://www.1000genomes.org/wiki/analysis/variant-call-format/bcf-binary-vcf-version-2`).

See also

 ▶ The specification for VCF is available at `http://samtools.github.io/hts-specs/VCFv4.2.pdf`

 ▶ GATK is one of the most widely used variant callers; check `https://www.broadinstitute.org/gatk/`

 ▶ samtools and htslib are used for variant calling and SAM/BAM management; check `http://htslib.org`

Studying genome accessibility and filtering SNP data

While previous recipes were focused on giving an overview of Python libraries to deal with alignment and variant call data, we concentrate on actually using them with a clear purpose in mind here.

If you are using NGS data, chances are that your most important file to analyze is a VCF file, produced by a genotype caller such as samtools mpileup, or GATK. The quality of your VCF calls may need to be assessed and filtered. Here, we will put in place a framework to filter SNP data. Rather than giving you filtering rules (an impossible task to be performed in a general way), we give you procedures to assess the quality of your data. With this, you can devise your own filters.

Getting ready

In the best-case scenario, you have a VCF file with proper filters applied. If this is the case, you can just go ahead and use your file. Note that all VCF files will have a FILTER column, but this might not mean that all proper filters were applied. You have to be sure that your data is properly filtered.

In the second case, which is one of the most common, your file will have unfiltered data, but you have enough annotations and can apply hard filters (there is no need for programmatic filtering). If you have a GATK annotated file, refer to `http://gatkforums.broadinstitute.org/discussion/2806/howto-apply-hard-filters-to-a-call-set`.

In the third case, you have a VCF file that has all the annotations that you need, but you may want to apply more flexible filters (for example, "if read depth > 20, accept if mapping quality > 30; otherwise, accept if mapping quality > 40").

In the fourth case, your VCF file does not have all the necessary annotations and you have to revisit your BAM files (or even other sources of information). In this case, the best solution is to find whatever extra information you need and create a new VCF file with required annotations. Some genotype callers (such as GATK) allow you to specify which annotations you want; you may also want to use extra programs to provide more annotations. For example, SnpEff (`http://snpeff.sourceforge.net/`) will annotate your SNPs with predictions of their effect (for example, if they are in exons, are they coding on noncoding?).

It's impossible to provide a clear-cut recipe. It will vary with the type of your sequencing data, your species of study, and your tolerance to errors, among other variables. What we can do is provide a set of typical analysis that is done for high-quality filtering.

In this recipe, we will not use data from the human 1000 genomes project. We want "dirty" unfiltered data that has a lot of common annotations that can be used to filter it. We will use data from the Anopheles 1000 genomes project (Anopheles is the mosquito vector involved in the transmission of the parasite that causes malaria), which makes filtered and unfiltered data available. You can find more information on this project at `http://www.malariagen.net/projects/vector/ag1000g`.

We will get a part of the centromere of chromosome 3L for around 100 mosquitoes, which is followed by a part somewhere in the middle of this chromosome (and index both):

```
tabix -fh
ftp://ngs.sanger.ac.uk/production/ag1000g/phase1/preview/ag1000g.
AC.phase1.AR1.vcf.gz  3L:1-200000 |bgzip -c > centro.vcf.gz
tabix -fh ftp://ngs.sanger.ac.uk/production/ag1000g/phase1/preview/
ag1000g.
AC.phase1.AR1.vcf.gz  3L:21000001-21200000 |bgzip -c >
standard.vcf.gz
tabix -p vcf centro.vcf.gz
tabix -p vcf standard.vcf.gz
```

If the links do not work, be sure to check `https://github.com/tiagoantao/bioinf-python/blob/master/notebooks/Datasets.ipynb` for updates. As usual, the code to download this data is available on a notebook: `01_NGS/Filtering_SNPs.ipynb`.

Finally, a word of warning on this recipe: the level of Python here will be slightly more complicated than usual. The more general code we will write will be easier to reuse in your specific case. We will use functional programming techniques (lambda functions) and the partial function application extensively.

How to do it...

Take a look at the following steps:

1. Let's start by plotting the distribution of variants across the genome in both files:

```python
%matplotlib inline
from collections import defaultdict

import seaborn as sns
import matplotlib.pyplot as plt

import vcf

def do_window(recs, size, fun):
    start = None
    win_res = []
    for rec in recs:
        if not rec.is_snp or len(rec.ALT) > 1:
            continue
        if start is None:
            start = rec.POS
        my_win = 1 + (rec.POS - start) // size
        while len(win_res) < my_win:
            win_res.append([])
        win_res[my_win - 1].extend(fun(rec))
    return win_res

wins = {}
size = 2000
vcf_names = ['centro.vcf.gz', 'standard.vcf.gz']
for vcf_name in vcf_names:
    recs = vcf.Reader(filename=vcf_name)
    wins[name] = do_window(recs, size, lambda x: [1])
```

- ❑ We start by performing the required imports (as usual, remember to remove the first line if you are not on the IPython Notebook). Before I explain the function, note what we are doing:

- ❑ For both files, we will compute windowed statistics. We divide our file which includes 200,000 bp of data in windows of size 2,000 (100 windows). Every time we find a biallelic SNP, we add a one to the list related to this window in the window function. The window function will take a VCF record (a SNP— `rec.is_snp`—that is not biallelic `len(rec.ALT) == 1`), determine the window where that record belongs (by performing an integer division of `rec.POS` by size), and extend the list of results of that window by the function passed to it as the fun parameter (which in our case is just a one).

- ❑ So, now we have a list of 100 elements (each representing 2,000 base pairs). Each element will be another list that will have a one for each biallelic SNP found. So, if you have 200 SNPs in the first 2,000 base pairs, the first element of the list will have 200 ones.

2. Let's continue:

```
def apply_win_funs(wins, funs):
    fun_results = []
    for win in wins:
        my_funs = {}
        for name, fun in funs.items():
            try:
                my_funs[name] = fun(win)
            except:
                my_funs[name] = None
        fun_results.append(my_funs)
    return fun_results


stats = {}
fig, ax = plt.subplots(figsize=(16, 9))
for name, nwins in wins.items():
    stats[name] = apply_win_funs(nwins, {'sum': sum})
    x_lim = [i * size  for i in range(len(stats[name]))]
    ax.plot(x_lim, [x['sum'] for x in stats[name]], label=name)
ax.legend()
ax.set_xlabel('Genomic location in the downloaded segment')
ax.set_ylabel('Number of variant sites (bi-allelic SNPs)')
fig.suptitle('Distribution of MQ0 along the genome', fontsize='xx-
large')
```

❑ Here, we perform a plot that contains statistical information for each of our 100 windows. The `apply_win_funs` will calculate a set of statistics for every window. In this case, it will sum all the numbers in the window. Remember that every time we find a SNP, we add a one to the window list. This means that if we have 200 SNPs, we will have 200 1s; hence, summing them will return 200.

❑ So, we are able to compute the number of SNPs per window in an apparently convoluted way. Why we perform things with this strategy will become apparent soon. However, for now, let's check the result of this computation for both files, as shown in the following figure:

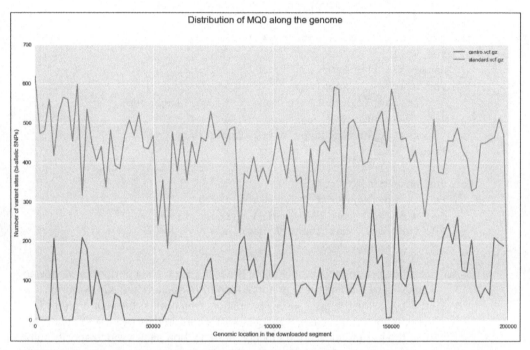

Figure 5: The number of bi-allelic SNPs distributed of windows of 2,000 bp of size for an area of 200 kbp near the centromere (blue) and in the middle of chromosome (green); both areas come from chromosome 3L for circa 100 Ugandan mosquitoes from the Anopheles 1000 genomes project

 Note that the amount of SNPs in the centromere is smaller than in the middle of the chromosome. This is expected because both calling variants in chromosomes are more difficult than calling in the middle. Also, there is probably less genomic diversity in centromeres. If you are used to humans or other mammals, you will find the density of variants obnoxiously high, that is mosquitoes for you!

3. Let's take a look at the sample-level annotation. We will inspect mapping quality zero (refer to `https://www.broadinstitute.org/gatk/guide/` `tooldocs/org_broadinstitute_gatk_tools_walkers_annotator_` `MappingQualityZeroBySample.php` for details), which is a measure of how well sequences involved in calling this variant map clearly to this position. Note that there is also a `MQ0` annotation at the variant-level:

```
import functools

import numpy as np
mq0_wins = {}
vcf_names = ['centro.vcf.gz', 'standard.vcf.gz']
size = 5000
def get_sample(rec, annot, my_type):
    res = []
    samples = rec.samples
    for sample in samples:
        if sample[annot] is None:  # ignoring nones
            continue
        res.append(my_type(sample[annot]))
    return res

for vcf_name in vcf_names:
    recs = vcf.Reader(filename=vcf_name)
    mq0_wins[vcf_name] = do_window(recs, size,
        functools.partial(get_sample,
            annot='MQ0', my_type=int))
```

   ❑ Start inspecting this by looking at the last `for`; we will perform a windowed analysis by reading the `MQ0` annotation from each record. We perform this by calling the `get_sample` function, which will return our preferred annotation (in this case, `MQ0`) cast with a certain type (`my_type=int`). We use the partial application function here. Python allows you to specify some parameters of a function and wait for other parameters to be specified later. Note that the most complicated thing here is the functional programming style. Also, note that it makes it very easy to compute other sample-level annotations. Just replace `MQ0` with AB, AD, GQ, and so on. You immediately have a computation for that annotation. If the annotation is not of type integer, no problem; just adapt `my_type`. It's a difficult programming style if you are not used to it, but you will reap its benefits very soon.

4. Let's now print the median and top 75 percent percentile for each window (in this case, with a size of 5,000):

```
stats = {}
colors = ['b', 'g']
i = 0
```

```
fig, ax = plt.subplots(figsize=(16, 9))
for name, nwins in mq0_wins.items():
    stats[name] = apply_win_funs(nwins, {'median':
np.median, '75': functools.partial(np.percentile, q=75)})
    x_lim = [j * size  for j in range(len(stats[name]))]
    ax.plot(x_lim, [x['median'] for x in stats[name]], label=name,
        color=colors[i])
    ax.plot(x_lim, [x['75'] for x in stats[name]], '--',
        color=colors[i])
    i += 1
ax.legend()
ax.set_xlabel('Genomic location in the downloaded segment')
ax.set_ylabel('MQ0')
fig.suptitle('Distribution of MQ0 along the genome',
    fontsize='xx-large')
```

❑ Note that we now have two different statistics on `apply_win_funs` (percentile and median). Again, we pass functions as parameters (`np.median` and `np.percentile`) with partial-function application done on `np.percentile`. The result is as follows:

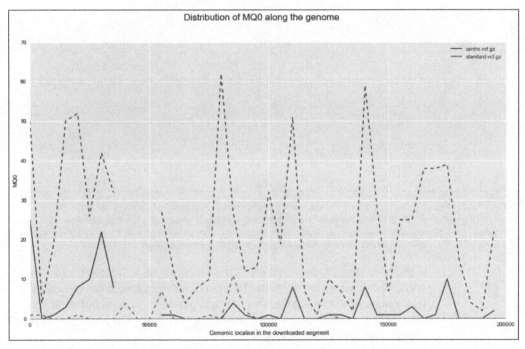

Figure 6: Median (continuous line) and 75th percentile (dashed) of MQ0 of sample SNPs distributed on windows of 5,000 bp of size for an area of 200 kbp near the centromere (blue) and in the middle of chromosome (green); both areas come from chromosome 3L for circa 100 Ugandan mosquitoes from the Anopheles 1000 genomes project

❑ For the "standard" file, the median MQ0 is 0 (it's plotted at the very bottom and is almost unseen). This is good as it suggests that most sequences involved in the calling of variants map clearly to this area of the genome. For the centromere, MQ0 is of poor quality. Furthermore, there are areas where the genotype caller will not find any variants at all (hence, the incomplete chart).

5. Let's compare heterozygosity with (DP), the sample-level annotation. Here, we will plot the fraction of heterozygosity calls as a function of the sample read depth (DP) for every SNP. We first explain the result and then the code that generates it.

❑ The following figure shows the fraction of calls that are heterozygous at a certain depth:

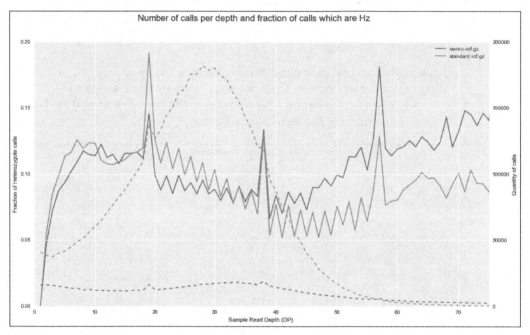

Figure 7: The continuous line represents the fraction of heterozygosite calls computed at a certain depth; in blue is the centromeric area, in green is the "standard" area; the dashed lines represent the number of sample calls per depth; both areas come from chromosome 3L for circa 100 Ugandan mosquitoes from the Anopheles 1000 genomes project

❑ In the preceding figure, there are two considerations to be taken into account. At a very low depth, the fraction of heterozygote calls is biased low. This makes sense as the number of reads per position does not allow you to make a correct estimate of the presence of both alleles in a sample. So, you should not trust calls at very low depth.

- As expected, the number of calls in the centromere is way lower than outside it. The distribution of SNPs outside the centromere follows a common pattern that you can expect in many datasets. The code is as follows:

```
def get_sample_relation(recs, f1, f2):
    rel = defaultdict(int)
    for rec in recs:
        if not rec.is_snp:
            continue
        for sample in rec.samples:
            try:
                v1 = f1(sample)
                v2 = f2(sample)
                if v1 is None or v2 is None:
                    continue  # We ignore Nones
                rel[(v1, v2)] += 1
            except:
                pass # This is outside the domain
    return rel


rels = {}
for vcf_name in vcf_names:
    recs = vcf.Reader(filename=vcf_name)
    rels[vcf_name] = get_sample_relation(recs, lambda s: 1
        if s.is_het else 0, lambda s: int(s['DP']))
```

- Start by looking at the `for` loop. Again, we use functional programming; the `get_sample_relation` function will traverse all SNP records and apply two functional parameters. The first parameter determines heterozygosity, whereas the second parameter acquires the sample DP (remember that there is also a variant DP):

- Now, as the code is as complex as it is, I opted for a naive data structure to be returned by `get_sample_relation:`, a dictionary where the key is a pair of results (in this case, heterozygosity and DP) and the sum of SNPs that share both values. There are more elegant data structures with different trade-offs. For this, SciPy spare matrices, pandas' DataFrames, or you may want to consider PyTables. The fundamental point here is to have a framework that is general enough to compute relationships between a couple of sample annotations.

- Also, be careful with the dimension space of several annotations. For example, if your annotation is of the float type, you might have to round it (if not, the size of your data structure might become too big).

6. Now, let's take a look at the plotting code. Let's perform this in two parts. Here is part one:

```
def plot_hz_rel(dps, ax, ax2, name, rel):
    frac_hz = []
    cnt_dp = []
    for dp in dps:
        hz = 0.0
        cnt = 0

        for khz, kdp in rel.keys():
            if kdp != dp:
                continue
            cnt += rel[(khz, dp)]
            if khz == 1:
                hz += rel[(khz, dp)]
        frac_hz.append(hz / cnt)
        cnt_dp.append(cnt)
    ax.plot(dps, frac_hz, label=name)
    ax2.plot(dps, cnt_dp, '--', label=name)
```

❑ This function will take a data structure as generated by `get_sample_relation`, expecting that the first parameter of the key tuple is the heterozygosity state (0 = homozygote, 1 = heterozygote) and the second parameter is the DP. With this, it will generate two lines: one with the fraction of samples (which are heterozygotes at a certain depth) and the other with the SNP count.

7. Now, let's call this function:

```
fig, ax = plt.subplots(figsize=(16, 9))
ax2 = ax.twinx()
for name, rel in rels.items():
    dps = list(set([x[1] for x in rel.keys()]))
    dps.sort()
    plot_hz_rel(dps, ax, ax2, name, rel)
ax.set_xlim(0, 75)
ax.set_ylim(0, 0.2)
ax2.set_ylabel('Quantity of calls')
ax.set_ylabel('Fraction of Heterozygote calls')
ax.set_xlabel('Sample Read Depth (DP)')
ax.legend()
fig.suptitle('Number of calls per depth and fraction of
calls which are Hz',,
            fontsize='xx-large')
```

❑ Here, we will use two axes. On the left-hand side, we will have the fraction of heterozygozite SNPs. On the right-hand side, we will have the number of SNPs. We then call `plot_hz_rel` for both data files. The rest is standard matplotlib code.

8. Finally, let's compare the variant DP with a categorical variant level annotation (EFF). EFF is provided by SnpEFF and tells us (among many other things) the type of SNP (for example, intergenic, intronic, coding synonymous, and coding nonsynonymous). The Anopheles dataset provides this useful annotation. Let's start by extracting variant-level annotations and the functional programming style:

```python
def get_variant_relation(recs, f1, f2):
    rel = defaultdict(int)
    for rec in recs:
        if not rec.is_snp:
            continue
        try:
            v1 = f1(rec)
            v2 = f2(rec)
            if v1 is None or v2 is None:
                continue  # We ignore Nones
            rel[(v1, v2)] += 1
        except:
            pass
    return rel
```

- The programming style here is similar to `get_sample_relation`, but we do not delve into samples. Now, we define the types of effects that we work with and convert its effect to an integer (as it will allow us to use it as in index, for example, matrices). Now, think about coding a categorical variable:

```python
accepted_eff = ['INTERGENIC', 'INTRON',
    'NON_SYNONYMOUS_CODING', 'SYNONYMOUS_CODING']

def eff_to_int(rec):
    try:
        for annot in rec.INFO['EFF']:
            #We use the first annotation
            master_type = annot.split('(')[0]
            return accepted_eff.index(master_type)
    except ValueError:
        return len(accepted_eff)
```

9. We now traverse the file; the style should be clear to you now:

```python
eff_mq0s = {}
for vcf_name in vcf_names:
    recs = vcf.Reader(filename=vcf_name)
    eff_mq0s[vcf_name] = get_variant_relation(recs, lambda
        r: eff_to_int(r), lambda r: int(r.INFO['DP']))
```

10. Finally, we plot the distribution of DP using the SNP effect:

```
fig, ax = plt.subplots(figsize=(16,9))
vcf_name = 'standard.vcf.gz'
bp_vals = [[] for x in range(len(accepted_eff) + 1)]
for k, cnt in eff_mq0s[vcf_name].items():
    my_eff, mq0 = k
    bp_vals[my_eff].extend([mq0] * cnt)
sns.boxplot(bp_vals, sym='', ax=ax)
ax.set_xticklabels(accepted_eff + ['OTHER'])
ax.set_ylabel('DP (variant)')
fig.suptitle('Distribution of variant DP per SNP type',
             fontsize='xx-large')
```

❑ Here, we just print a boxplot for the noncentromeric file, as shown in the following figure. The results are as expected: SNPs in coding areas will probably have more depth because they are in more complex regions that are easier to call than intergenic SNPs:

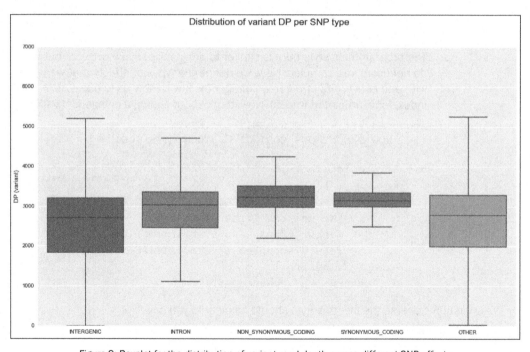

Figure 8: Boxplot for the distribution of variant read depth across different SNP effects

There's more...

The whole issue of filtering SNPs and other genome features will need a book on its own. The approach will depend on the type of sequencing data that you have, number of samples, and potential extra information (for example, pedigree among samples).

This recipe is very complex as it is, but parts of it are profoundly naive (there is a limit of complexity that I can force on you in a simple recipe). For example, the window code does not support overlapping windows. Also, data structures are simplistic. However, I hope that they give you an idea of the general strategy to process genomic high-throughput sequencing data.

See also

- There are many filtering rules, but I would like to draw your attention to the need of a reasonably good coverage (clearly above 10x). Refer to Meynet et al *Variant detection sensitivity and biases in whole genome and exome sequencing* at `http://www.biomedcentral.com/1471-2105/15/247/`.

- Brad Chapman is one of the best known specialists in sequencing analysis and data quality with Python and the main author of Blue Collar Bioinformatics, a blog that you might want to refer to at `https://bcbio.wordpress.com/`.

- Brad is also the main author of bcbio-nextgen a Python-based pipeline for high-throughput sequencing analysis (`https://bcbio-nextgen.readthedocs.org`).

- Peter Cock is the main author of Biopython and is heavily involved in NGS analysis. Be sure to check his blog, *Blasted Bioinformatics!?* at `http://blastedbio.blogspot.co.uk/`.

3
Working with Genomes

In this chapter, we will cover the following recipes:

- ▸ Working with high-quality reference genomes
- ▸ Dealing with low-quality reference genomes
- ▸ Traversing genome annotations
- ▸ Extracting genes from a reference using annotations
- ▸ Finding orthologues using the Ensembl REST API
- ▸ Retrieving gene ontology information from Ensembl

Introduction

Many tasks in computational biology are dependent on the existence of reference genomes. If you are performing a sequence alignment, finding genes, or studying genetics of populations at several points of your work, you will be directly or indirectly using a genome reference. In this chapter, we will develop some recipes to work with reference genomes and deal with varying quality of references (which can vary for high-quality, like with the human genome, to problematic with non-model species). We will also see how to deal with genome annotations (working with text databases that will point us to interesting features in the genome) and extract sequence data using the annotation information. Also, we will try to find some gene orthologues across species. Finally, we will access a **gene ontology** (**GO**) database.

Working with high-quality reference genomes

In this recipe, you will learn a few general techniques to manipulate reference genomes. As an illustrative example, we will study the GC content (the fraction of the genome that is based on Guanine-Cytosine). Reference genomes are normally made available as FASTA files.

Getting ready

Genomes come in widely different sizes, ranging from viruses such as HIV (which is 9.7 kbp) to bacteria such as *E. coli*, to protozoans such as *Plasmodium falciparum* (the most important parasite species causing malaria) with its 14 chromosomes, mitochondrion, and apicoplast, to the fruit fly with three autosomes, a mitochondrion, and X/Y sex chromosomes, to humans with its three Gbp pairs spread across 22 autosomes, X/Y chromosomes, and mitochondria, all the way up to *Paris japonica*, a plant with 150 Gbp of genome. Along the way, you have different ploidy and different sex chromosome organizations.

As you can see, different organisms have very different genome sizes. This difference can be of several orders of magnitude. This can have significant implications for your programming style. Working with a large genome will require you to be more conservative with the usage of memory. Unfortunately, larger genomes also benefit from more speed-efficient programming techniques (as you have much more data to analyze); these are conflicting requirements. The general rule is that you have to be much more careful with efficiency (both speed and memory) with larger genomes.

In order to make this recipe less burdensome, we will use a small eukaryotic genome from *Plasmodium falciparum*. This genome still has many typical features of larger genomes (for example, multiple chromosomes). So, it's a good compromise between complexity and size. Note that with a genome of the size of *P. falciparum*, it will be possible to perform many operations by loading the whole genome in-memory. However, we opted for a programming style that can be used with bigger genomes (for example, mammals) so that you can use this recipe in a more general way, but feel free to use more memory-intensive approaches with small genomes like this.

We will use Biopython, which you installed in *Chapter 1*, *Python and the Surrounding Software Ecology*. As usual, this recipe is available in the IPython Notebook at `02_Genomes/ Reference_Genome.ipynb` in the code bundle of the book.

If you are not using notebooks, download the *P. falciparum* genome from our datasets page at `https://github.com/tiagoantao/bioinf-python/blob/master/notebooks/ Datasets.ipynb` (file `pfalciparum.fasta`)

How to do it...

Let's take a look at the following steps:

1. We start by inspecting the description of all the sequences on the reference genome FASTA file:

```
from Bio import SeqIO
genome_name = 'PlasmoDB-9.3_Pfalciparum3D7_Genome.fasta'
recs = SeqIO.parse(genome_name, 'fasta')
for rec in recs:
    print(rec.description)
```

 ❑ This code should look familiar from the previous chapter; let's take a look at a part of the output:

```
Pf3D7_05_v3 | organism=Plasmodium_falciparum_3D7 | version=2012-02-01 | length=1343557 | SO=chromosome
Pf3D7_10_v3 | organism=Plasmodium_falciparum_3D7 | version=2012-02-01 | length=1687656 | SO=chromosome
Pf3D7_07_v3 | organism=Plasmodium_falciparum_3D7 | version=2012-02-01 | length=1445207 | SO=chromosome
Pf3D7_03_v3 | organism=Plasmodium_falciparum_3D7 | version=2012-02-01 | length=1067971 | SO=chromosome
Pf3D7_13_v3 | organism=Plasmodium_falciparum_3D7 | version=2012-02-01 | length=2925236 | SO=chromosome
Pf3D7_11_v3 | organism=Plasmodium_falciparum_3D7 | version=2012-02-01 | length=2038340 | SO=chromosome
Pf3D7_14_v3 | organism=Plasmodium_falciparum_3D7 | version=2012-02-01 | length=3291936 | SO=chromosome
Pf3D7_09_v3 | organism=Plasmodium_falciparum_3D7 | version=2012-02-01 | length=1541735 | SO=chromosome
Pf3D7_01_v3 | organism=Plasmodium_falciparum_3D7 | version=2012-02-01 | length=640851 | SO=chromosome
Pf3D7_12_v3 | organism=Plasmodium_falciparum_3D7 | version=2012-02-01 | length=2271494 | SO=chromosome
Pf3D7_08_v3 | organism=Plasmodium_falciparum_3D7 | version=2012-02-01 | length=1472805 | SO=chromosome
Pf3D7_06_v3 | organism=Plasmodium_falciparum_3D7 | version=2012-02-01 | length=1418242 | SO=chromosome
Pf3D7_04_v3 | organism=Plasmodium_falciparum_3D7 | version=2012-02-01 | length=1200490 | SO=chromosome
Pf3D7_02_v3 | organism=Plasmodium_falciparum_3D7 | version=2012-02-01 | length=947102 | SO=chromosome
M76611 | organism=Plasmodium_falciparum_3D7 | version=2012-02-01 | length=5967 | SO=mitochondrial_chromosome
PFC10_API_IRAB | organism=Plasmodium_falciparum_3D7 | version=2012-02-01 | length=34242 | SO=apicoplast_chromosome
```

 ❑ Different genome references will have different description lines, but they will generally have important information over there. In this example, you can see that we have chromosomes, mitochondria, and apicoplast. We can also view chromosome sizes, but we will take the value from the sequence length instead.

2. Let's parse the description line to extract the chromosome number. We will retrieve the chromosome size from the sequence and compute the GC content across chromosomes on a window basis:

```
from __future__ import print_function
from Bio import SeqUtils

recs = SeqIO.parse(genome_name, 'fasta')
chrom_sizes = {}
chrom_GC = {}
block_size = 50000
min_GC = 100.0
max_GC = 0.0
for rec in recs:
    if rec.description.find('SO=chromosome') == -1:
```

```
        continue
chrom = int(rec.description.split('_')[1])
chrom_GC[chrom] = []
size = len(rec.seq)
chrom_sizes[chrom] = size
num_blocks = size // block_size + 1
for block in range(num_blocks):
    start = block_size * block
    if block == num_blocks - 1:
        end = size
    else:
        end = block_size + start + 1
    block_seq = rec.seq[start:end]
    block_GC = SeqUtils.GC(block_seq)
    if block_GC < min_GC:
        min_GC = block_GC
    if block_GC > max_GC:
        max_GC = block_GC
    chrom_GC[chrom].append(block_GC)
print(min_GC, max_GC)
```

❑ We perform a windowed analysis of all chromosomes, similar to what you have seen in the previous chapter. We start by defining a window size of 50 kbp. This is appropriate for *P. falciparum* (feel free to vary its size), but you will want to consider other values for genomes with chromosomes that are orders of magnitude different from this.

❑ Note that we are re-reading the file. With such a small genome, it would have been feasible (in step one) to do an in-memory load of the whole genome. By all means, feel free to try this programming style for small genomes—it's faster! However, this code is more generalized for larger genomes.

❑ Note that in the `for` loop, we ignore the mitochondrion and apicoplast by parsing the SO entry to the description. The `chrom_sizes` dictionary will maintain the size of chromosomes.

❑ The `chrom_GC` dictionary is our most interesting data structure and will have a list of faction of the GC content for each 50 kbp window. So, for chromosome 1, which has a size of 640,851 bp, there will be 14 entries because this chromosome size has 14 blocks of 50 kbp.

 Be aware of two unusual features of the *P. falciparum* genome: the genome is very AT-rich, that is, GC-poor. So, the numbers that you will get will be very low. Also, chromosomes are ordered based on size (as it's common), but starting with the smallest size. The usual convention is to start with the largest size (for example, like genomes in humans).

3. Now, let's perform a genome plot of the GC distribution. We will use shades of blue for the GC content. However, for high outliers, we will use shades of red. For low outliers, we will use shades of yellow:

```
from __future__ import division
from reportlab.lib import colors
from reportlab.lib.units import cm
from Bio.Graphics import BasicChromosome
```

❑ We will use float division and import functions required by Biopython from the `reportlab` library:

The Biopython code has evolved over time, before Python was such a fashionable language. In the past, availability of libraries was quite limited. The usage of `reportlab` can be seen mostly as a legacy issue. I suggest that you learn just enough from it to use it with Biopython. If you are planning on learning a modern plotting library in Python, you will probably want to consider matplotlib, Bokeh, or Python's version of ggplot (or other visualization alternatives, such as Mayavi, VTK, or even Blender's API).

```
chroms = list(chrom_sizes.keys())
chroms.sort()
biggest_chrom = max(chrom_sizes.values())
my_genome = BasicChromosome.Organism(output_format='png')
my_genome.page_size = (29.7*cm, 21*cm)
telomere_length = 10
bottom_GC = 17.5
top_GC = 22.0
for chrom in chroms:
    chrom_size = chrom_sizes[chrom]
    chrom_representation = BasicChromosome.Chromosome \
        ('Cr %d' % chrom)
    chrom_representation.scale_num = biggest_chrom
    tel = BasicChromosome.TelomereSegment()
    tel.scale = telomere_length
    chrom_representation.add(tel)
    num_blocks = len(chrom_GC[chrom])
    for block, gc in enumerate(chrom_GC[chrom]):
        my_GC = chrom_GC[chrom][block]
        body = BasicChromosome.ChromosomeSegment()
        if my_GC > top_GC:
            body.fill_color = colors.Color(1, 0, 0)
        elif my_GC < bottom_GC:
            body.fill_color = colors.Color(1, 1, 0)
```

```
        else:
            my_color = (my_GC - bottom_GC) / (top_GC -
            bottom_GC)
            body.fill_color = colors.Color(my_color,
            my_color, 1)
        if block < num_blocks - 1:
            body.scale = block_size
        else:
            body.scale = chrom_size % block_size
        chrom_representation.add(body)
    tel = BasicChromosome.TelomereSegment(inverted=True)
    tel.scale = telomere_length
    chrom_representation.add(tel)
    my_genome.add(chrom_representation)
my_genome.draw('falciparum.png', 'Plasmodium falciparum')
```

- ❏ The first line converts the return of the keys method to a list. This is redundant in Python 2, but not in Python 3, where the keys method has a specific return type: `dict_keys`.

- ❏ We will draw the chromosomes in order (hence the sort). We will need the size of the biggest chromosome (14 in *P. falciparum*) in order to assure that the size of chromosomes is printed with the correct scale (the `biggest_chrom` variable).

- ❏ We then create an A4-sized representation of an organism with a PNG output. Note that we will draw very small telomeres of 10 bp. This will produce a rectangular-like chromosome. You can make the telomeres bigger, giving it a roundish representation (or you may have a better idea of the correct telomere size for your species).

- ❏ We declare that anything with GC content below 17.5 percent or above 22.0 percent will be considered an outlier. Remember that for most other species, this will be much higher.

- ❏ We then print these chromosomes proper. They are bounded by telomeres and composed of 50 kbp chromosome segments (the last segment is sized with the remainder). Each segment will be colored in blue with a red-green component based on the linear normalization between two outlier values. Each chromosome segment will either be 50 kbp or potentially smaller if the last one of the chromosome. The output is shown in the following figure:

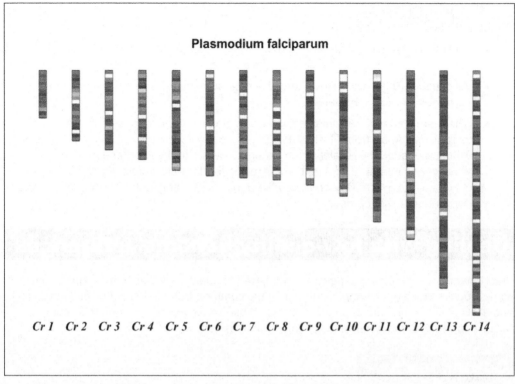

Figure 1: The 14 chromosomes of Plasmodium falciparum color-coded with the GC content (red is more than 22 percent, yellow less than 17 percent, and the blue shades represent a linear gradient between both numbers

4. Finally, if you are on the IPython Notebook (that is, do not perform this outside IPython), you can print the image inline:

```
from IPython.core.display import Image
Image("falciparum.png")
```

There's more...

P. falciparum is a reasonable example for a eukaryote with a small genome that allows you to perform a small data exercise with enough features that make it still useful for most eukaryotes. Of course, there are no sex chromosomes (such as X/Y in humans), but these should be easy to process because reference genomes do not deal with ploidy issues.

P. falciparum does have a mitochondrion, but we will not deal with it here due to space issues. Biopython does have the functionality to print circular genomes, which you can also use with bacteria. With regards to bacteria and viruses, these genomes are much easier to process because their size is very small.

See also

▸ You can find many reference genomes of model organisms in Ensembl at `http://www.ensembl.org/info/data/ftp/index.html`.

▸ As usual, NCBI also provides a large list of genomes at `http://www.ncbi.nlm.nih.gov/genome/browse/`.

▸ There are plenty of websites dedicated to a single organism (or a set of related organisms). Apart from PlasmoDB (`http://plasmodb.org/plasmo/`), from which you downloaded the *P. falciparum* genome, you will find VectorBase (`https://www.vectorbase.org/`) in the next recipe for disease vectors. Flybase (`http://flybase.org/`) for *Drosophila* is also worth mentioning, but do not forget to search for your organism of interest.

Dealing with low-quality genome references

Unfortunately, not all reference genomes will have the quality of *P. falciparum*. Apart from some model species (for example, humans or the common fruit fly *Drosophila melanogaster*) and a few others, most reference genomes could use some improvement. In this recipe, we will see how to deal with reference genomes with less quality.

Getting ready

In keeping with the malaria theme, here, we will use the reference genomes of two mosquitoes that are vectors of malaria: *Anopheles gambiae* (which is the most important vector of malaria and can be found in sub-Saharan Africa) and *Anopheles atroparvus*, a malaria vector in Europe (while the disease has been eradicated in Europe, this vector is still around). The *An. gambiae* genome is of reasonable quality. Most chromosomes have been mapped, although the Y chromosome still needs some work. There is a fairly large "unknown" chromosome, probably composed of bits X and Y chromosomes and also midgut microbiota. This genome has a reasonable amount of positions that are not called (that is, you will find Ns instead of ACTGs). The *An. atroparvus* genome is still in the scaffold format. Unfortunately, this is what you will find for many non-model species.

Note that we will up the ante a bit. The Anopheles genome is one order of magnitude bigger than the *P. falciparum* genome (but one order of magnitude smaller than most mammals).

We will use Biopython, which you installed in *Chapter 1, Python and the Surrounding Software Ecology*. As usual, this recipe is available from the IPython Notebook at `02_Genomes/Low_Quality.ipynb` in the code bundle of the book.

If you are not using notebooks, download the Anopheles genomes from our dataset page at `https://github.com/tiagoantao/bioinf-python/blob/master/notebooks/Datasets.ipynb` (files `gambiae.fa.gz` and `atroparvus.fa.gz`). Rename the first as `gambiae.fa.gz` and the second as `atroparvus.fa.gz`.

How to do it...

Let's take a look at the following steps:

1. Let's start by listing the chromosomes of the *A. gambiae* genome:

```python
from __future__ import print_function
import gzip
from Bio import SeqIO
gambiae_name = 'gambiae.fa.gz'
atroparvus_name = 'atroparvus.fa.gz'
recs = SeqIO.parse(gzip.open(gambiae_name), 'fasta')
for rec in recs:
    print(rec.description)
```

   - This will produce the following output:

```
chromosome:AgamP3:2L:1:49364325:1 chromosome 2L
chromosome:AgamP3:2R:1:61545105:1 chromosome 2R
chromosome:AgamP3:3L:1:41963435:1 chromosome 3L
chromosome:AgamP3:3R:1:53200684:1 chromosome 3R
chromosome:AgamP3:UNKN:1:42389979:1 chromosome UNKN
chromosome:AgamP3:X:1:24393108:1 chromosome X
chromosome:AgamP3:Y_unplaced:1:237045:1 chromosome Y_unplaced
```

   - The code is quite straightforward. We use the `gzip` module because files of larger genomes are normally compressed. We can see four chromosome arms (2L, 2R, 3L, and 3R) and the X chromosome. Note the Y chromosome, which is quite small and has a name that all but indicates that it might not be in the best state. Also, note that the unknown (UNKN) chromosome is actually a quite large proportion of the reference genome to the tune of the chromosome arm.

   - Do not perform this with *An. atroparvus* or you will get more than a thousand entries, courtesy of the scaffold status.

2. We will now check uncalled positions (Ns) and its distribution for the *An. gambiae* genome:

```python
from __future__ import division
recs = SeqIO.parse(gzip.open(gambiae_name), 'fasta')
chrom_Ns = {}
chrom_sizes = {}
```

```
for rec in recs:
    chrom = rec.description.split(':')[2]
    if chrom in ['UNKN', 'Y_unplaced']:
        continue
    chrom_Ns[chrom] = []
    on_N = False
    curr_size = 0
    for pos, nuc in enumerate(rec.seq):
        if nuc in ['N', 'n']:
            curr_size += 1
            on_N = True
        else:
            if on_N:
                chrom_Ns[chrom].append(curr_size)
                curr_size = 0
            on_N = False
    if on_N:
        chrom_Ns[chrom].append(curr_size)
    chrom_sizes[chrom] = len(rec.seq)
for chrom, Ns in chrom_Ns.items():
    size = chrom_sizes[chrom]
    print('%s (%s): %%Ns (%.1f), num Ns: %d, max N: %d' % (
        chrom, size, 100 * sum(Ns) / size, len(Ns), max(Ns)))
```

❑ This code will take some time to run, so please be patient because we will inspect each and every base pair of the autosomes. As usual, we will reopen and re-read the file to save memory.

❑ We have two dictionaries: one dictionary with chromosome sizes and another with the distribution of the sizes of runs of Ns. To calculate the runs of Ns, we traverse all autosomes (noting when an N position starts and ends). We then print the basic statistics of the distribution of Ns:

```
2L (49364325): %Ns (1.7), num Ns: 957, max N: 28884
3R (53200684): %Ns (1.8), num Ns: 1128, max N: 24292
X (24393108): %Ns (4.1), num Ns: 1287, max N: 21132
2R (61545105): %Ns (2.3), num Ns: 1658, max N: 36427
3L (41963435): %Ns (2.9), num Ns: 1272, max N: 31063
```

❑ So, for the 2L chromosome arm (with a size of 49 Mbp), 1.7 percent are N calls divided by 957 runs. The biggest run is 28,884 bps. Note that the X chromosome has the highest fraction of positions with Ns.

3. We will now turn our attention to the *An. Atroparvus* genome. Let's count the number of scaffolds along with the distribution of scaffold sizes:

```
import numpy as np
recs = SeqIO.parse(gzip.open(atroparvus_name), 'fasta')
sizes = []
size_N = []
for rec in recs:
    size = len(rec.seq)
    sizes.append(size)
    count_N = 0
    for nuc in rec.seq:
        if nuc in ['n', 'N']:
            count_N += 1
    size_N.append((size, count_N / size))

print(len(sizes), np.median(sizes), np.mean(sizes),
        max(sizes), min(sizes),
        np.percentile(sizes, 10), np.percentile(sizes, 90))
```

□ This code is similar to the previous point, but we do print slightly more detailed statistics using NumPy, so we get:

```
(1371, 8304.0, 163596.0, 20238125, 1004, 1563.0, 56612.0)
```

□ We thus have 1,371 scaffolds (against seven entries on the *A. gambiae* genome) with a median size of 8,304 (mean of 163,536). The biggest scaffold has 2 Mbp and the smallest scaffold has 1004 bp. The tenth percentile for size is 1,563 and the ninetieth is 56,612.

4. Finally, let's plot the fraction of the scaffold, that is, N as a function of its size. If on IPython, prepend `%matplotlib` inline to the following code:

```
import matplotlib.pyplot as plt

small_split = 4800
large_split = 540000
fig, axs = plt.subplots(1, 3, figsize=(16, 9),
    squeeze=False)
xs, ys = zip(*[(x, 100 * y) for x, y in size_N
            if x <= small_split])
axs[0, 0].plot(xs, ys, '.')
axs[0, 0].set_ylim(-0.1, 3.5)
xs, ys = zip(*[(x, 100 * y) for x, y in size_N
   if x > small_split and x <= large_split])
axs[0, 1].plot(xs, ys, '.')
axs[0, 1].set_xlim(small_split, large_split)
xs, ys = zip(*[(x, 100 * y) for x, y in size_N
```

```
          if x > large_split])
axs[0, 2].plot(xs, ys, '.')
axs[0, 0].set_ylabel('Fraction of Ns', fontsize=12)
axs[0, 1].set_xlabel('Contig size', fontsize=12)
fig.suptitle('Fraction of Ns per contig size',
      fontsize=26)
```

❑ The preceding code will generate the output shown in the following figure, in which we split the chart into three parts based on the scaffold size: one for scaffolds with less than 4,800 bp, one for scaffolds between 4,800 and 540,000 bp, and one for larger ones. Note that the *y* axis scale (depicting the fraction of the N scaffold) is completely different for the three panels. It is very low for small scaffolds (always below 44.5) with large variance (between 0 and 100 percent), for medium scaffolds and tighter variance (between 0 and 25 percent) for the largest scaffold:

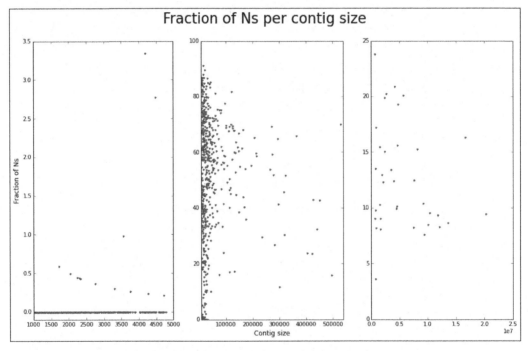

Figure 2: Fraction of scaffolds that are N as a function of their size

There's more...

Sometimes, reference genomes carry extra information, for example, the *Anopheles gambiae* genome is soft masked. This means that some procedures were run on the genome to identify areas of low complexity (which are normally more problematic to analyze). This will be annotated by capitalization: ACTG will be high complexity, whereas ACTG will be low.

Reference genomes with lots of scaffolds are more than an inconvenient hassle. For example, very small scaffolds (say, below 2,000 bp) may have mapping problems when using an aligner (such as BWA), especially at the extremes (most scaffolds will have mapping problems at their extremes, but these will be a much larger proportion of the scaffold if it's small). If you are using a reference genome like this to align, you will want to consider ignoring the pair information (assuming that you have paired-end reads) when mapping to small scaffolds, or at least measure the impact of the scaffold size in the performance of your aligner. In any case, the general comment is that be careful because the scaffold size and number will rear its ugly head from time to time.

With these genomes, there was only complete ambiguity (N) identified. Note that other genome assemblies will give you an intermediate code between total ambiguity and certainty (ACTG).

See also

▸ Tools like RepeatMasker are used to find areas of the genome with low complexity at `http://www.repeatmasker.org/`

▸ IUPAC ambiguity codes may be useful to have in hand when processing other genomes at `http://www.bioinformatics.org/sms/iupac.html`

Traversing genome annotations

Having a genome sequence is interesting, but we will want to extract features from it: genes, exons, and coding sequences. This type of annotation information is made available in GFF and GTF files. GFF stands for Generic Feature Format. In this recipe, we will see how to parse and analyze GFF files, using the annotation of the *Anopheles gambiae* genome as an example.

Getting ready

We will use the `gffutils` library to process the annotation file.

If you do not use the notebook, you need to acquire the annotation file from our datasets page at `https://github.com/tiagoantao/bioinf-python/blob/master/notebooks/Datasets.ipynb` (file `gambiae.gff3.gz`) Rename the annotation file as `gambiae.gff.gz`. Preferably, use the `02_Genomes/Annotations.ipynb` notebook, which is provided in the code bundle of the book.

How to do it...

Let's take a look at the following steps:

1. Let's start by creating an annotation database with `gffutils` based on our GFF file:

```
import gffutils
import sqlite3
try:
    db = gffutils.create_db('gambiae.gff.gz', 'ag.db')
except sqlite3.OperationalError:
    db = gffutils.FeatureDB('ag.db')
```

   ❑ The `gffutils` library creates a SQLite database to store annotations efficiently. Here, we try to create the database, but if it already exists, we will use the existing one. This step can be time-consuming.

2. Now, we list all the available feature types and count them:

```
print(list(db.featuretypes()))
for feat_type in db.featuretypes():
    print(feat_type, db.count_features_of_type(feat_type))
```

   ❑ Features will include contigs, genes, exons, transcripts, and so on. Note that we will use the `gffutils` package's `featuretypes` function. It will return a generator, but we will convert it to a list (it's safe here).

3. Let's list all contigs:

```
for contig in db.features_of_type('contig'):
    print(contig)
```

   ❑ This will show that there is an annotation information for all chromosome arms and sex chromosomes, mitochondrion, and the unknown chromosome.

4. Let's now extract a lot of useful information per chromosome, number of genes, number of transcripts per gene, number of exons, and so on:

```
from collections import defaultdict
num_mRNAs = defaultdict(int)
num_exons = defaultdict(int)
max_exons = 0
max_span = 0
for contig in db.features_of_type('contig'):
    cnt = 0
    for gene in db.region((contig.seqid, contig.start,
contig.end), featuretype='gene'):
        cnt += 1
        span = abs(gene.start - gene.end) # strand
```

```
        if span > max_span:
            max_span = span
            max_span_gene = gene
        my_mRNAs = list(db.children(gene,
featuretype='mRNA'))
        num_mRNAs[len(my_mRNAs)] += 1
        if len(my_mRNAs) == 0:  # checking num of
alternative transcripts
            exon_check = [gene]
        else:
            exon_check = my_mRNAs
        for check in exon_check:
            my_exons = list(db.children(check,
            featuretype='exon'))  # level = SLOW
            num_exons[len(my_exons)] += 1
            if len(my_exons) > max_exons:
                max_exons = len(my_exons)
                max_exons_gene = gene
    print('contig %s, number of genes %d' % (contig.seqid,
        cnt))
print('Max number of exons: %s (%d)' % (max_exons_gene.id,
    max_exons))
print('Max span: %s (%d)' % (max_span_gene.id, max_span))
print(num_mRNAs)
print(num_exons)
```

- ❑ We traverse all contigs (`features_of_type`), extracting all genes (`region`). In each gene, we count the number of alternative transcripts. If there are none (note that this is probably an annotation issue and not a biological one), we count the exons (children). If there are several transcripts, we count the exons per transcript. We also account for the span size to check for the gene spanning the largest region. We follow a similar procedure to find the gene and the largest number of exons.

- ❑ Finally, we print a dictionary with the distribution of the number of alternative transcripts per gene (`num_mRNAs`) and the distribution of number of exons per transcript (`num_exons`).

There's more...

There are many variations of the GFF/GTF format. There are different GFF versions and many unofficial variations. If possible, choose the GFF Version 3. However, the ugly truth is that you will find it very difficult to process files. The `gffutils` library tries as best as it can to accommodate this. Indeed, much of the documentation of the library is concerned with helping you process all kinds of awkward variations (refer to `https://pythonhosted.org/gffutils/examples.html`).

There is an alternative to using `gffutils` (either because your GFF file is strange or because you do not like the library interface or its dependency on a SQL backend). Parse the file yourself manually. If you look at the format, you will notice that it's not very complex. If you are only performing a one-off operation, then maybe manual parsing is good enough. Of course, one-off operations tend to not be that.

Also, note that the quality of annotations tend to vary a lot. As the quality increases, so does the complexity. Just check the human annotation for an example of this. One can expect that over time, as our knowledge of organisms evolves, the quality and complexity of annotations will increase.

See also

- ▶ The GFF spec can be found at `https://www.sanger.ac.uk/resources/software/gff/spec.html`.

- ▶ Probably the best explanation on the GFF format, along with the most common versions and GTF, can be found at `http://gmod.org/wiki/GFF3`.

- ▶ Somewhat related is the BED format to specify annotations; this is mostly used interactively to specify tracks for genome viewers. You will probably come across it at `http://genome.ucsc.edu/FAQ/FAQformat.html#format1`, although processing is quite trivial.

Extracting genes from a reference using annotations

In this recipe, we will see how to extract a gene sequence with the help of an annotation file to get its coordinates against a reference FASTA. We will use the *Anopheles gambiae* genome, along with its annotation file (as per the previous two recipes). We will first extract the **Voltage-gated sodium channel** (**VGSC**) gene, which is involved in resistance to insecticides.

Getting ready

If you have followed the previous two recipes, you are ready. If not, download the *Anopheles gambiae* FASTA file, along with the GTF file. You also need to prepare the `gffutils` database:

```
import gffutils
import sqlite3
try:
    db = gffutils.create_db('gambiae.gff.gz', 'ag.db')
except sqlite3.OperationalError:
    db = gffutils.FeatureDB('ag.db')
```

As usual, you will find all this in the `02_Genomes/Getting_Gene.ipynb` notebook.

How to do it...

Let's take a look at the following steps:

1. Let's start by retrieving the annotation information for our gene:

```
import gzip
from Bio import Alphabet, Seq, SeqIO
gene_id = 'AGAP004707'
gene = db[gene_id]
print(gene)
print(gene.seqid, gene.strand)
```

   ❑ The gene ID was retrieved from VectorBase, an online database of genomics of disease vectors. For other specific cases, you will need to know the ID of your gene (which will be dependent on species and database). The output will be as follows:

```
2L        VectorBase     gene     2358158 2431617 .        +        .        ID=AGAP004707;biotype=protein_coding
('2L', '+')
```

   ❑ Note that the gene is on the 2L chromosome arm and coded in the positive direction.

2. Let's hold the sequence for the 2L chromosome arm in memory (it's just a single chromosome, so we will indulge):

```
recs = SeqIO.parse(gzip.open('gambiae.fa.gz'), 'fasta')
for rec in recs:
    print(rec.description)
    if rec.description.split(':')[2] == gene.seqid:
        my_seq = rec.seq
        break
print(my_seq.alphabet)
```

   ❑ The output will be as follows:

```
chromosome:AgamP3:2L:1:49364325:1 chromosome 2L
SingleLetterAlphabet()
```

   ❑ Note the alphabet of the sequence.

3. Let's create a function to construct a gene sequence for a list of CDSs:

```
def get_sequence(chrom_seq, CDSs, strand):
    seq = Seq.Seq('',
        alphabet=Alphabet.IUPAC.unambiguous_dna)
    for CDS in CDSs:
```

```
        my_cds = Seq.Seq(str(my_seq[CDS.start - 1:
CDS.end]), alphabet=Alphabet.IUPAC.unambiguous_dna)
        seq += my_cds
    return seq if strand == '+' else
seq.reverse_complement()
```

- ❏ This function will receive a chromosome sequence (in our case, the 2L arm), a list of coding sequences (retrieved from the annotation file), and the strand.

- ❏ There are several important issues to note here. We will construct a sequence of the unambiguous DNA type (we will need this for translation). The `SingleLetterAlphabet` of the previous step will have to be converted. We also have to be very careful with the start and end of the sequence (note that the GFF file is 1-based, whereas the Python array is 0-based). Finally, we return the reverse complement if the strand is negative.

4. Although we have the gene ID at hand, we actually want to just one of the available transcripts of the three available for this gene, so we need to choose one:

```
mRNAs = db.children(gene, featuretype='mRNA')
for mRNA in mRNAs:
    print(mRNA.id)
    if mRNA.id.endswith('RA'):
        break
```

5. We will now get the coding sequence for our transcript and get the gene sequence and translate it:

```
CDSs = db.children(mRNA, featuretype='CDS',
    order_by='start')
gene_seq = get_sequence(my_seq, CDSs, gene.strand)

print(len(gene_seq), gene_seq)
prot = gene_seq.translate()
print(len(prot), prot)
```

6. Let's get the gene that is coded in the negative strand direction. We will just take the gene next to VGSC (which happens to be the negative strand):

```
reverse_transcript_id = 'AGAP004708-RA'
reverse_CDSs = db.children(reverse_transcript_id,
    featuretype='CDS', order_by='start')
reverse_seq = get_sequence(my_seq, reverse_CDSs, '-')

print(len(reverse_seq), reverse_seq)
reverse_prot = reverse_seq.translate()
print(len(reverse_prot), reverse_prot)
```

Here, I evaded getting all the information about the gene and just hardcoded the transcript ID. The point is that you should make sure your code works irrespective of strand.

There's more...

This is a simple recipe that exercises several concepts that have been presented in this and the previous chapter. While it's conceptually trivial, it's unfortunately full of booby traps.

 When using different databases, be sure that the genome assembly versions are in synchrony. It would be a serious and potentially silent bug to use different versions. Remember that different versions (at least on the major version number) have different coordinates. For example, human position 1234 on chromosome 3 on build 36 will probably refer a different SNP than 1234 on build 38. With human data, you will probably find a lot of chips on build 36, plenty of whole genome sequences on build 37, whereas the most recent human assembly is build 38. With our Anopheles example, you will have versions 3 and 4 around. This will happen with most species. So, be aware!

There is also the issue of 0-indexed arrays in Python versus 1-indexed genomic databases. Be nonetheless aware that some genomic databases may also be 0-indexed.

There are also two sources of confusion: the transcript versus the gene choice as in more rich annotation databases, you will have several alternative transcripts (if you want to look at the rich-to-the-point-of-confusing database, refer to the human annotation database). Also, fields tagged with exon will have more information than the coding sequence. For this purpose, you will want the CDS field.

Finally, there is the strand issue where you will want to translate based on the reverse complement.

See also

- You can download MySQL tables for Ensembl at http://www.ensembl.org/info/data/mysql.html.

- The UCSC genome browser can be found at http://genome.ucsc.edu/. Be sure to check the download area at http://hgdownload.soe.ucsc.edu/downloads.html.

- As with reference genomes, you can find GTFs of model organisms in Ensembl at http://www.ensembl.org/info/data/ftp/index.html.

- A simple explanation between CDSs and exons can be found at https://www.biostars.org/p/65162/.

Finding orthologues with the Ensembl REST API

Here, we will see how to look for orthologues for a certain gene. This simple recipe will not only introduce orthology retrieval, but also how to use REST APIs on the Web to access biological data. Last but surely not least, it will serve as an introduction on how to access the Ensembl database using the programmatic API.

In our example, we will try to find any orthologue for the human lactase gene (LCT) on the horse genome.

Getting ready

This recipe will not require any pre-downloaded data, but as we are using the web APIs, Internet access will be needed. The amount of data transferred will be limited.

We will also make use of the requests library to access Ensembl. The request API is an easy-to-use wrapper for web requests. Of course, you can use the standard Python libraries, but these are much more cumbersome.

As usual, you can find this content on the `02_Genomes/Orthology.ipynb` notebook.

How to do it...

Let's take a look at the following steps:

1. We start by creating a support function to perform a web request:

```
import requests

ensembl_server = 'http://rest.ensembl.org'

def do_request(server, service, *args, **kwargs):
    url_params = ''
    for a in args:
        if a is not None:
            url_params += '/' + a
    req = requests.get('%s/%s%s' % (server, service,
url_params),
                       params=kwargs,
                       headers={'Content-Type':
'application/json'})
    if not req.ok:
        req.raise_for_status()
    return req.json()
```

    ❑   We start by importing the requests library and specifying the root URL. Then, we create a simple function that will take the functionality to be called (see the following examples) and generate a complete URL. It will also add optional parameters and specify the payload to be of the JSON type (just to get a default JSON answer). It will return the response in the JSON format. This is typically a nested Python data structure of lists and dictionaries.

2. We start by checking all the available species on the server, which is around 40 at the time of this writing:

```
all_species = do_request(ensembl_server, 'info/species')
for sp in all_species['species']:
    print(sp['name'])
```

    ❑   Note that this will construct the `http://rest.ensembl.org/info/species` URL for the REST request.

3. We will now try to find any HGNC databases on the server related to human data:

```
ext_dbs = do_request(ensembl_server, 'info/external_dbs',
                'homo_sapiens', filter='HGNC%')
print(ext_dbs)
```

    ❑   We restrict the search to human-related databases (`homo_sapiens`). We also filter databases starting with HGNC (this filtering uses the SQL notation). HGNC is the HUGO database. We want to make sure that it's available because the HUGO database is responsible for curating human gene names and maintains our LCT identifier.

4. Now that we know that the LCT identifier is probably available, we want to retrieve the Ensembl ID for the gene, as shown in the following code:

```
ensembl = do_request(ensembl_server, 'lookup/symbol',
                'homo_sapiens', 'LCT')
print(ensembl)
lct_id = ensembl['id']
```

 Different databases, as you probably know by now, will have different IDs for the same object. We will need to resolve our LCT identifier to the Ensembl ID. When you deal with external databases relating same objects, ID translation among databases will probably be your first task.

5. Just for your information, we can now get the sequence of the area containing the gene. Note that this is probably the whole interval, and if you want to recover the gene, you will have to use a procedure similar to the previous recipe:

```
lct_seq = do_request(ensembl_server, 'sequence/id', lct_id)
print(lct_seq)
```

6. We can also inspect other databases that are known to Ensembl to refer to this gene:

```
lct_xrefs = do_request(ensembl_server, 'xrefs/id', lct_id)
for xref in lct_xrefs:
    print(xref['db_display_name'])
    print(xref)
```

   ❑ You will find different kinds of databases, such as vertebrate and genome annotation project (VEGA), UniProt (see *Chapter 7, Using the Protein Data Bank*), or WikiGene.

7. Finally, let's get the orthologues for this gene on the horse genome:

```
hom_response = do_request(ensembl_server, 'homology/id',
    lct_id, type='orthologues', sequence='none')
homologies = hom_response['data'][0]['homologies']
for homology in homologies:
    print(homology['target']['species'])
    if homology['target']['species'] != 'equus_caballus':
        continue
    print(homology)
    print(homology['taxonomy_level'])
    horse_id = homology['target']['id']
```

   ❑ We could have actually acquired the orthologues directly for the horse by specifying a `target_species` parameter. However, this code allows you to inspect all available orthologues.

   ❑ You will get quite a lot of information about an orthologue, such as the taxonomy level of orthology (Boreoeutheria—placental mammals is the closest phylogenetic level between humans and horses), the ensembl ID of the orthologue, the dn/ds ratio (nonsynonymous to synonymous mutations), and the CIGAR string (refer to the previous chapter) of differences among sequences. By default, you will also get the alignment of the orthologous sequence, but I have removed it to unclog the output.

8. Finally, let's look for the horse ID Ensembl record:

```
horse_req = do_request(ensembl_server, 'lookup/id',
    horse_id)
print(horse_req)
```

From this point onwards, you can use the previous recipe methods to explore the LCT horse orthologue.

There's more...

You can find a detailed explanation of all the functionalities available at `http://rest.ensembl.org/`. This includes all the interfaces and Python code snippets, among other languages.

If you are interested in paralogues, this information can be retrieved quite trivially from the preceding recipe. On the call to `homology/id`, just replace the type with paralogues.

If you have heard of Ensembl, you have probably heard of an alternative service from UCSC: the Genome Browser (`http://genome.ucsc.edu/`). While on a user interface, they are on the same level, from a programmatic perspective, Ensembl is probably more mature. Access to NCBI Entrez databases was covered in the previous chapter.

Another completely different strategy to interface programmatically with Ensembl will be to download raw tables and inject them into a local MySQL database. Be aware that this will be quite an undertaking in itself (you will probably just want to load a very small subset of tables). However, if you intend to be very intensive in terms of usage, you may have to consider creating a local version of part of the database. If this is the case, you may want to reconsider the UCSC alternative as it's as good as Ensembl from the local database perspective.

Retrieving gene ontology information from Ensembl

In this recipe, we will introduce the usage of gene ontology information again by querying the Ensembl REST API. Gene ontologies are controlled vocabularies to annotate gene and gene products. These are made available as trees of concepts (with more general concepts near the top of the hierarchy). There are three domains for gene ontologies: cellular component, molecular function, and biological process.

Getting ready

As with the previous recipe, we do not require any predownloaded data, but as we are using web APIs, Internet access will be needed. The amount of data transferred will be limited.

As usual, you can find this content on the `02_Genomes/Gene_Ontology.ipynb` notebook.

We will make use of the `do_request` function, which is defined in the first step of the previous recipe (Finding Orthologues with the REST Ensembl API).

To draw GO trees, we will use `pygraphviz`, a graph drawing library.

How to do it...

Let's take a look at the following steps:

1. Let's start by retrieving all GO terms associated with the LCT gene (you can see how to retrieve the Ensembl ID in the previous recipe). Remember that you will need the `do_request` function from the previous recipe:

```
lct_id = 'ENSG00000115850'
refs = do_request(ensembl_server, 'xrefs/id', lct_id,
    external_db='GO', all_levels='1')
print(len(refs))
print(refs[0].keys())
for ref in refs:
    go_id = ref['primary_id']
    details = do_request(ensembl_server, 'ontology/id',
        go_id)
    print('%s %s %s' % (go_id,  details['namespace'],
        ref['description']))
    print('%s\n' % details['definition'])
```

   ❑ Note the free-form definition and the varying namespace for each term. The first two of the eleven reported items in the loop are (this may change when you run it because the database may have been updated):

```
GO:0000016 molecular_function lactase activity
"Catalysis of the reaction: lactose + H2O = D-glucose + D-galactose." [EC:3.2.1.108]

GO:0004553 molecular_function hydrolase activity, hydrolyzing O-glycosyl compounds
"Catalysis of the hydrolysis of any O-glycosyl bond." [GOC:mah]
```

2. Let's concentrate on the "lactase activity" molecular function and retrieve more detailed information about it (the following GO ID comes from the previous step):

```
go_id = 'GO:0000016'
my_data = do_request(ensembl_server, 'ontology/id', go_id)
for k, v in my_data.items():
    if k == 'parents':
        for parent in v:
            print(parent)
            parent_id = parent['accession']
    else:
        print('%s: %s' % (k, str(v)))
parent_data = do_request(ensembl_server, 'ontology/id',
    parent_id)
print(parent_id, len(parent_data['children']))
```

 ❑ We print the "lactase activity" record (which is currently a node of the GO tree molecular function) and retrieve a list of potential parents. There is a single parent for this record. We retrieve it and print the number of children.

3. Let's retrieve all the general terms for the "lactase activity" molecular function (again, the parent and all other ancestors):

```
refs = do_request(ensembl_server, 'ontology/ancestors/chart',
    go_id)
for go, entry in refs.items():
    print(go)
    term = entry['term']
    print('%s %s' % (term['name'], term['definition']))
    is_a = entry.get('is_a', [])
    print('\t is a: %s\n' % ', '.join([x['accession'] for x
    in is_a]))
```

 ❑ We retrieve the ancestor list by following the is_a relationship (refer to the following GO sites for more details on the types of possible relationships).

4. Let's define a function to create a dictionary with the ancestor relationship for a term along with some summary information for each term returned in a pair:

```
def get_upper(go_id):
    parents = {}
    node_data = {}
    refs = do_request(ensembl_server,
        'ontology/ancestors/chart', go_id)
    for ref, entry in refs.items():
        my_data = do_request(ensembl_server, 'ontology/id',
            ref)
        node_data[ref] = {'name': entry['term']['name'],
        'children': my_data['children']}
        try:
            parents[ref] = [x['accession'] for x in
                entry['is_a']]
        except KeyError:
            pass  # Top of hierarchy
    return parents, node_data
```

5. We will finally print a tree of relationships for the "lactase activity" term. For this, we will use the `pygraphivz` library:

```
parents, node_data = get_upper(go_id)
import pygraphviz as pgv
g = pgv.AGraph(directed=True)
for ofs, ofs_parents in parents.items():
    ofs_text = '%s\n(%s)' % (node_data[ofs]['name'].replace(', ',
        '\n'), ofs)
    for parent in ofs_parents:
        parent_text = '%s\n(%s)' % (node_data[parent]['name'].
            replace(', ', '\n'), parent)
        children = node_data[parent]['children']
        if len(children) < 3:
            for child in children:
                if child['accession'] in node_data:
                    continue
                g.add_edge(parent_text, child['accession'])
        else:
            g.add_edge(parent_text, '...%d...' % (len(children) -
1))
        g.add_edge(parent_text, ofs_text)
print(g)
g.graph_attr['label']='Ontology tree for Lactase activity'
g.node_attr['shape']='rectangle'
g.layout(prog='dot')
g.draw('graph.png')
```

   ❏ The following figure shows the ontology tree for the "lactase activity" term. Terms at the top are more general. The top of the tree denotes `molecular_function`. For all ancestral nodes, the number of extra offspring is also noted (or enumerated if less than three).

6. Of course, if you are on IPython, you can inline the image using the following code:

```
from IPython.core.display import Image
Image("graph.png")
```

❑ The output is shown here:

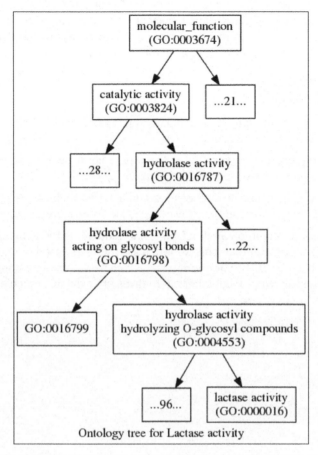

Figure 3: An ontology tree for the term "lactase activity" (terms at the top are more general);
the top of the tree is molecular_function; for all ancestral nodes, the number of extra offspring is also
noted (or enumerated if less than three)

There's more...

If you are interested in gene ontologies, your main port of call will be http://
geneontology.org, where you will find much more information on this topic. Apart
from molecular_function, gene ontology also has a "biological process" and a "cellular
component". In our recipes, we have followed the hierarchical relationship "is a", but others do
exist partially. For example, "mitochondrial ribosome" (GO:0005761) is a cellular component,
a part of "mitochondrial matrix" (refer to http://amigo.geneontology.org/amigo/
term/GO:0005761#display-lineage-tab and click on the "graph view").

As with the previous recipe, you can download the MySQL dump of a gene ontology database (you may prefer to interact with the data in that way). For this, see `http://geneontology.org/page/go-database`. Again, expect to allocate some time to understand the relational database schema. Also, note that there are many alternative ways to graphviz in order to plot trees and graphs. We will return to this topic in the course of this book.

See also

- As stated before, more than Ensembl, the main resource for gene ontologies is `http://geneontology.org`.

- For visualization, we are using the `pygraphviz` library, which is a wrapper on top of graphviz (`http://www.graphviz.org`).

- There are very good user interfaces for GO data, for example, AmiGO (`http://amigo.geneontology.org`) and QuickGO (`http://www.ebi.ac.uk/QuickGO/`).

- One of the most common analyses performed with GO is the gene enrichment analysis to check whether some GO terms are over expressed or under expressed in a certain gene set. The `geneontology.org` server uses Panther (`http://go.pantherdb.org/`), but other alternatives are available (such as DAVID) (`http://david.abcc.ncifcrf.gov/`).

4
Population Genetics

In this chapter, we will cover the following recipes:

- ▶ Managing datasets with PLINK
- ▶ Introducing the Genepop format
- ▶ Exploring a dataset with Bio.PopGen
- ▶ Computing F-statistics
- ▶ Performing Principal Components Analysis
- ▶ Investigating population structure with Admixture

Introduction

Population genetics is the study of changes of frequency of alleles in a population on the basis of selection, drift, mutation, and migration. The previous chapters focused mainly on data processing and cleanup; this is the first chapter in which we will actually infer interesting biological results.

There is a lot of interesting population genetics analysis based on sequence data, but as we already have quite a few recipes to deal with sequence data, we will divert our attention somewhere else. Also, we will not cover genomic structural variation such as **Copy Number Variation** (**CNVs**) or inversions here. I will concentrate on analyzing SNP data, which is one of the most common. We will perform many standard analyses, including population genetic analyses with Python, such as F_{ST} (**Fixation index**), **Principal Components Analysis** (**PCA**), and study of population structure.

We will use Python as a scripting language that glues together applications that perform necessary computations, which is the "old-fashioned way". Having said that, as the Python software ecology is still evolving, you can at least perform the PCA in Python using **scikit-learn.** Also, we will perform the PCA using EIGENSOFT's **smartpca** (an external application).

There is no such thing as a default file format for population genetics data. The bleak reality of this field is that there are plenty of formats, most of them developed with a specific application in mind; therefore, it's not generically applicable. Some of the efforts to create a more general format (or even just a file converter to support many formats) met with limited success. Furthermore, as our knowledge of genomics increases, we will require new formats anyway (for example, to support some kind of previously unknown genomic structural variation). Here, we will work with the formats of two widely used applications. One is PLINK (`http://pngu.mgh.harvard.edu/~purcell/plink/`).This format was originally developed to perform genome-wide association studies (GWAS) with human data and has many more applications. The second format is Genepop (`http://kimura.univ-montp2.fr/~rousset/Genepop.htm`). This format is widely used in the conservation genetics community. If you have NGS sequencing data, you may question why not VCF? Well, a VCF file is normally annotated to help with sequencing analysis, which you do not need at this stage (you should now have a filtered dataset). If you convert your SNP calls from VCF to PLINK, you will gain 95 percent in terms of size (this is in comparison to a compressed VCF). More importantly, the computational cost of processing a VCF file is much bigger (think of processing all this highly-structured text) than the cost of other two formats.

This chapter is closely tied to the next one. Here, we will work with real empirical data, whereas in the next chapter, we will simulate data. However, if you are interested in analyzing population genetics data, be sure to read the next chapter, where we will cover some ground analysis.

First, let's start with a discussion on file format issues and then continue to discuss interesting data analysis.

Managing datasets with PLINK

Here, we will manage our dataset using PLINK. We will create subsets of our main dataset (from the HapMap project) suitable to be analyzed in our next recipes.

Note that neither PLINK nor any similar programs were developed because of their file formats. Probably, it had no objective to become a default file standard for population genetics data. In this field, you will need to be ready to convert from format to format (for this, Python is quite appropriate) because every application that you will use will probably have its own quirky requirements. The most important point to learn from this recipe is that it's not formats that are being used, although these are relevant, but the "file conversion mentality". Apart from this, some of the steps in this recipe also convey genuine analysis techniques that you may want to consider using (for example, subsampling or **Linkage Disequilibrium** (**LD**) pruning).

Getting ready

Throughout this chapter, we will use data from the Human HapMap project. You may recall that we used data from the Human 1000 genomes project in *Chapter 2, Next-generation Sequencing*, the HapMap project is in many ways the precursor of the Human 1000 genomes project; instead of whole genome sequencing, genotyping was used. Most of the samples of the HapMap project were used in the Human 1000 genomes project, so if you have read the recipes in *Chapter 2, Next-generation Sequencing*, you will already have an idea of the dataset (including the available population). I will not introduce the dataset much more, but you can refer to *Chapter 2, Next-generation Sequencing*, and, of course, the HapMap site (http://www.hapmap.org). Remember that we have genotyping data for many individuals split across populations around the globe. We will refer to these populations by their acronyms. Here is the list taken from http://www.sanger.ac.uk/resources/downloads/human/hapmap3.html:

Acronym	Population
ASW	This denotes African ancestry in Southwest USA
CEU	This denotes Utah residents with Northern and Western European ancestry from the CEPH collection
CHB	This denotes Han Chinese in Beijing, China
CHD	This denotes Chinese in Metropolitan Denver, Colorado
GIH	This denotes Gujarati Indians in Houston, Texas
JPT	This denotes Japanese in Tokyo, Japan
LWK	This denotes Luhya in Webuye, Kenya
MXL	This denotes Mexican ancestry in Los Angeles, California

Acronym	Population
MKK	This denotes Maasai in Kinyawa, Kenya
TSI	This denotes Toscani in Italia
YRI	This denotes Yoruba in Ibadan, Nigeria

This will require a fairly big download (approximately 1 GB), which will have to be uncompressed. For this, you will need `bzip2`, which is available at `http://www.bzip.org/`. Make sure that you have approximately 20 GB of disk space for this chapter.

You can download this from `https://github.com/tiagoantao/bioinf-python/ blob/master/notebooks/Datasets.ipynb`, the `hapmap.map.bz2` and `hapmap.ped.bz2` files, plus the relationships.txt file

Decompress the PLINK file using the following commands:

```
bunzip2 hapmap3_r2_b36_fwd.consensus.qc.poly.map.bz2
bunzip2 hapmap3_r2_b36_fwd.consensus.qc.poly.ped.bz2
```

Therefore, we have PLINK files; the MAP file has positions of the information of the marker position across the genome, whereas the PED file has actual markers for each individual along with some pedigree information. We also downloaded a metadata file that contains information about each individual. Take a look at all these files and familiarize yourself with them.

As usual, this is also available in the `03_PopGen/Data_Formats.ipynb` notebook, where everything has been taken care of.

Finally, most of this recipe will make heavy usage of PLINK (you should have installed at least version 1.9 from `http://pngu.mgh.harvard.edu/~purcell/plink/` not version 1.0x). Python will mostly be used as the glue language to call PLINK.

How to do it...

Take a look at the following steps:

1. Let's get the metadata for our samples. We will load the population of each sample and note all individuals that are offspring of others in the dataset:

```
from collections import defaultdict
f = open('relationships_w_pops_121708.txt')
pop_ind = defaultdict(list)
f.readline()  # header
offspring = []
for l in f:
    toks = l.rstrip().split('\t')
    fam_id = toks[0]
    ind_id = toks[1]
```

```
    mom = toks[2]
    dad = toks[3]
    if mom != '0' or dad != '0':
        offspring.append((fam_id, ind_id))
    pop = toks[-1]
    pop_ind[pop].append((fam_id, ind_id))
f.close()
```

❏ This will load a dictionary where population is the key (CEU, YRI, and so on) and its value is the list of individuals in that population. This dictionary will also store if the individual is the offspring of another. Each individual is identified by the family and individual ID (which are found on the PLINK file). The file provided by the HapMap project is a simple tab-delimited file, which is not difficult to process.

❏ There is an important point to make here, that is, the reason this information is provided on a separate, ad hoc file is because the PLINK format makes no provision for the population structure (this format makes provision only for the case/control information for which PLINK was designed). This is not a flaw of the format in a sense that it was never designed to support standard population genetic studies (it's a GWAS tool). However, this is a general "feature" of population genetics data formats: whichever you end up working with, there will be something important missing.

❏ We will use this metadata in other recipes in this chapter. We will also perform some consistency analysis between the metadata and the PLINK file, but we will defer this to the next recipe.

❏ pandas will be a good alternative to implement similar functionalities to read metadata.

2. Let's now subsample the dataset at 10 percent and 1 percent of the number of markers as follows:

```
import os
os.system('plink --recode --file
hapmap3_r2_b36_fwd.consensus.qc.poly --noweb --out hapmap10
--thin 0.1 --geno 0.1')
os.system('plink --recode --file
hapmap3_r2_b36_fwd.consensus.qc.poly --noweb --out hapmap1
--thin 0.01 --geno 0.1')
```

❏ Alternatively, if you are using IPython, you can simply use the following command:

```
!plink --recode --file hapmap3_r2_b36_fwd.consensus.qc.poly
--noweb --out hapmap10 --thin 0.1 --geno 0.1
!plink --recode --file hapmap3_r2_b36_fwd.consensus.qc.poly
--noweb --out hapmap1 --thin 0.01 --geno 0.1
```

- ❑ Note the subtlety that you will not really get 1 or 10 percent of the data; each marker will have a 1 or 10 percent chance of being selected, so you will get approximately 1 or 10 percent of markers.

- ❑ Obviously, as the process is random, different runs will produce different marker subsets. This will have important implications further down the road. If you want to replicate the exact same result, you can nonetheless use the --seed option.

- ❑ We will also remove all SNPs that have a genotyping rate of lower than 90 percent (the --geno 0.1 parameter).

There is nothing special about Python in this code, but there are two reasons you may want to subsample your data. First, if you are performing exploratory analysis of your own dataset, you may want to start with a smaller version because it will be easy to process. Also, you will have a broader view of your data. Second, some analysis methods may not require all your data (indeed, some methods might not be even able to use all your data). Be very careful with the last point though, that is, for every method that you use to analyze your data, be sure that you understand the data requirements for the scientific questions you want to answer. Feeding too much data may be okay normally (you pay a time and memory penalty), but feeding too little will lead to unreliable results.

3. Now, let's generate subsets with just the autosomes (that is, let's remove sex chromosomes and the mitochondria) as follows:

```python
def get_non_auto_SNPs(map_file, exclude_file):
    f = open(map_file)
    w = open(exclude_file, 'w')
    for l in f:
        toks = l.rstrip().split('\t')
        chrom = int(toks[0])
        rs = toks[1]
        if chrom > 22:
            w.write('%s\n' % rs)
    w.close()
get_non_auto_SNPs('hapmap1.map', 'exclude1.txt')
get_non_auto_SNPs('hapmap10.map', 'exclude10.txt')
os.system('plink --recode --file hapmap1 --noweb --out
hapmap1_auto --exclude exclude1.txt')
os.system('plink --recode --file hapmap10 --noweb --out
hapmap10_auto --exclude exclude10.txt')
```

❑ Let's create a function that generates a list with all SNPs not belonging to autosomes. In PLINK, this means a chromosome number above 22 (for 22 human autosomes). If you use another species, be careful with your chromosome coding because PLINK is geared towards human data. If your species are diploid and have less than 23 autosomes and a sex determination system, that is, X/Y, this will be straightforward; if not, refer to https://www.cog-genomics.org/plink2/input#allow_extra_chr for some alternatives (similar to the --allow-extra-chr flag).

❑ We then create autosome only PLINK files for subsample datasets of 10 and 1 percent (prefixed as hapmap10_auto and hapmap1_auto).

4. Let's create some datasets without an offspring that will be needed for most population genetic analysis, which require unrelated individuals to a certain degree:

```
os.system('plink --file hapmap10_auto --filter-founders --recode --out hapmap10_auto_noofs')
```

This step is representative of the fact that most population genetic analysis require samples to be unrelated to a certain degree. Obviously, as we know that some offsprings are in HapMap, we remove them. However, note that with your dataset, you are expected to be much more refined than this, for instance, run plink --genome or use another program to detect related individuals. The fundamental point here is that you have to dedicate some effort to detect related individuals in your samples; this is not a trivial task.

5. We will also generate an LD pruned dataset, as required by many PCA and Admixture algorithms, as follows:

```
os.system('plink --file hapmap10_auto_noofs --indep-pairwise 50 10 0.1 --out keep')
os.system('plink --file hapmap10_auto_noofs --extract keep.prune.in --recode --out hapmap10_auto_noofs_ld')
```

❑ The first step generates a list of markers to be kept if the dataset is LD-pruned. This uses a sliding window of 50 SNPs, advancing by 10 SNPs at a time with a cut value of 0.1.

❑ The second step extracts SNPs from the list generated earlier.

6. Let's recode a couple of cases in different formats:

```
os.system('plink --file hapmap10_auto_noofs_ld --recode12 tab --out hapmap10_auto_noofs_ld_12')
os.system('plink --make-bed --file hapmap10_auto_noofs_ld --out hapmap10_auto_noofs_ld')
```

- ❏ The first operation will convert a PLINK format that uses nucleotide letters (ACTG) to another, which recodes alleles with 1 and 2. We will use this in the *Performing Principal Components Analysis* recipe.

- ❏ The second operation recodes a file in a binary format. If you work inside PLINK (using the many useful operations that PLINK has), the binary format is probably the most appropriate format (for example, smaller file size). We will use this in the Admixture recipe.

7. We will also extract a single chromosome (2) for analysis. We will start with the autosome dataset subsampled at 10 percent:

```
os.system('plink --recode --file hapmap10_auto_noofs --
chr 2 --out hapmap10_auto_noofs_2')
```

There's more...

There are many reasons why you might want to create different datasets for analysis. You may want to perform some fast initial exploration of data; the analysis algorithm that you plan to use has some data format requirements or a constraint on the input, it could be the number of markers or relationships among individuals. Chances are that you will have lots of subsets to analyze (unless your dataset is very small to start with, for instance, a microsatellite dataset).

This may seem a minor point, but it's not. Be very careful with file naming (note that I have followed some simple conventions while generating filenames). Make sure that the name of the file gives some information about subset options. When you perform the downstream analysis, you will want to be sure that you choose the correct dataset; you will want your dataset management to be agile and reliable above all. The worst thing that can happen is to create an analysis with an erroneous dataset that does not obey constraints required by software.

At the time of writing this book, there are two PLINK versions: 1.x and the fast approaching version 2. I strongly suggest that you use version 2 beta as well because the speed and memory improvements in version 2 are impressive.

The LD-pruning that we used is somewhat standard for human analysis, but be sure to check the parameters, especially if you are using nonhuman data.

The HapMap file that we downloaded is based on an old version of the reference genome (build 36). As stated in the previous chapter, be sure to use annotations from build 36 if you plan to use this file for more analysis of your own.

This recipe will set the stage for all the next recipes and its results will be used extensively.

See also

▶ The Wikipedia page `http://en.wikipedia.org/wiki/Linkage_disequilibrium` on Linkage Disequilibrium is a good place to start

▶ The website of PLINK `http://pngu.mgh.harvard.edu/~purcell/plink/` is perfectly documented (something that many genetics software lacks)

Introducing the Genepop format

The Genepop format is used in many conservation genetics studies. It's the format of the Genepop application and is the de facto format for many population genetics analysis. If you come from other fields (for example, have a lot of sequencing experience), you may not have heard of it, but this format is widely used (as its citation record proves) and is worth a look. Here, we will convert some datasets from previous recipes to this format and introduce the Genepop parser from Biopython.

Getting ready

You will need to run the previous recipe because its output will be required for this one.

If you are not using Docker, you might not be using some of the code that I produced earlier (mostly to deal with the bread and butter of data conversion). You can find this code at `https://github.com/tiagoantao/pygenomics` and install it from

```
pip install pygenomics
```

Note that at this stage, we will not use the Genepop application (this will change in the next recipe), so no need to install it for now.

As usual, this is available in the `03_PopGen/Genepop_Format.ipynb` notebook, but it will still require you to run the previous notebook in order to generate the required files.

How to do it...

Take a look at the following steps:

1. Let's load the metadata (we will use a simplified version from the previous recipe) as follows:

```
from collections import defaultdict
f = open('relationships_w_pops_121708.txt')
pop_ind = defaultdict(list)
f.readline()  # header
for line in f:
    toks = line.rstrip().split('\t')
```

```
        fam_id = toks[0]
        ind_id = toks[1]
        pop = toks[-1]
        pop_ind[pop].append((fam_id, ind_id))
    f.close()
```

2. Let's check for consistency between the PLINK data file and the metadata, as we will need to clean up population mappings to generate a Genepop file, as shown in the following code:

```
all_inds = []
for inds in pop_ind.values():
    all_inds.extend(inds)
for line in open('hapmap1.ped'):
    toks = line.rstrip().replace(' ', '\t').split('\t')
    fam = toks[0]
    ind = toks[1]
    if (fam, ind) not in all_inds:
        print('Problems with %s/%s' % (fam, ind))
```

- The preceding code generates a list with all individuals coming from the metadata file. Then, it will open hapmap1.ped, which has the pedigree information at 1 percent sampling. (I have chosen 1 percent because 1 will be much faster to process than 10 or 100 percent samples; we only need pedigree, not genetic information) and compare the information on both. It will report all individuals that are on the PED file, but not on the metadata file.

- We perform a replacement procedure on each PED line because you can find some PLINK files that are space-separated, whereas others are tab-separated.

- In a perfect world, this will output nothing, but there is one incorrect entry. This entry (which has family ID 2469 and individual ID NA20281) is not consistent with the family ID reported on the metadata.

For your own dataset, always be sure to thoroughly compare your data with your metadata and to check for consistency problems. With all your sources of data (if you have more than one), make sure that they are consistent among themselves. If not, at least annotate all problematic cases, better yet, take action to understand and correct any underlying problems. The default assumption should be that there are problems (not that everything is sound). Although you may have produced the data yourself, check it. Bugs and typos are assured to happen. Making errors is normal, and checking for them is fundamental. Being overconfident is a sign of inexperience.

3. Let's convert some datasets from PLINK to the Genepop format:

```
from genomics.popgen.plink.convert import to_genepop
to_genepop('hapmap1_auto', 'hapmap1_auto', pop_ind)
to_genepop('hapmap10', 'hapmap10', pop_ind)
to_genepop('hapmap10_auto', 'hapmap10_auto', pop_ind)
to_genepop('hapmap10_auto_noofs_ld',
    'hapmap10_auto_noofs_ld', pop_ind)
to_genepop('hapmap10_auto_noofs_2',
    'hapmap10_auto_noofs_2', pop_ind)
```

- ❑ Note that we will pass the prefix of all files, especially the first one, to the input file. This will be prepended with .ped and .map to find input files. The second one will be prepended with .gp to generate the Genepop file, whereas .pops will contain the order of populations on the Genepop file. Take a look at both generated files in order to be familiar with the content, although we will dissect the result a bit more in the next recipe.

- ❑ As PLINK has no population structure information, we need to pass the pop_ind dictionary. This dictionary will be used to create a Genepop file structured by population.

- ❑ This uses a function provided by my package to convert PLINK to Genepop data. This will take some time to run. Note that we are just converting subsampled data as this is done to make things computationally more efficient in downstream analysis, but be aware that in many of your own analysis, you may need the complete dataset. The function will ignore individuals without population, which means that it will exclude the individual with the wrong family ID detected on the consistency step. A will be converted to 1, C to 2, T to 3, and G to 4. Although a pops file will be produced with the order of populations on the output file, this will always be lexicographically ordered.

- ❑ If you are curious about how this function works, feel free to take a look at https://github.com/tiagoantao/pygenomics/blob/master/genomics/popgen/plink/convert.py. Be forewarned as it particularly contains text processing.

4. Biopython provides an in-memory parser for Genepop files; let's take a small taste of it by opening the autosome file sampled at 1 percent:

```
from Bio.PopGen.GenePop import read
rec = read(open('hapmap1_auto.gp'))
print('Number of loci %d' % len(rec.loci_list))
print('Number of populations %d' % len(rec.pop_list))
print('Population names: %s' % ', '.join(rec.pop_list))
print('Individuals per population %s' % ', '.join([str(len(inds))
    for inds in rec.populations]))
ind = rec.populations[1][0]
```

```
print('Individual %s, SNP %s, alleles: %d %d' % (ind[0],
    rec.loci_list[0], ind[1][0][0], ind[1][0][1]))
del rec
```

❑   The output is as follows:

```
Number of loci 13902
Number of populations 11
Population names: 2436/NA19983, 1459/NA12865, NA18594/NA18594, NA18140/NA18140, NA20881/NA20881, NA19007/NA19007, NA19372/NA
19372, M005/NA19652, 2581/NA21371, NA20757/NA20757, Y105/NA19099
Individuals per population 82, 165, 84, 85, 88, 86, 90, 77, 171, 88, 167
Individual 1328/NA06989, SNP 1/rs2710888/949705, alleles: 3 2
```

❑   The default assumption about population names on Genepop is that somehow the last individual is used to identify a population. As this is slightly ad hoc, we will also generate a .pop file (as in the previous recipe) with names of populations.

❑   As the marker sampling process in the previous recipe is stochastic, you will probably see a slightly different number of loci.

❑   As the whole dataset is in memory, we can directly access any individual of any population. This is what we perform to print the last line, that is, we access the first individual of the second population and print its name along with alleles of the first SNP (which are 3 and 2, thus coding T and C). The first SNP is called 1/rs2710888/949705. Here, 1 represents the chromosome number, the middle ID is the SNP RS ID (the identifier on NCBI's dbSNP database), and the last number is the chromosome position against the human build 36.

❑   At the end, we delete the record because it takes up a lot of memory.

Note that some of these outputs depend on how the Genepop was coded (on my to_genepop function) and are based on that. For example, the coding of ACTG to 1234 is arbitrary (just a convenience) or the fact that populations are lexicographically ordered or loci names include the rs id and position chromosomes. If you receive your files from another source, you will have to check whatever conventions they have used (which may or may not be convenient to you). If you generate your own files, be sure to use conventions that will be useful downstream (like here). Of course, this argument is generalizable; you can apply it to other file formats as long as they have any form of built-in flexibility.

5.   More realistically, we will use the large file parser for most modern datasets because it won't load the whole in-memory file, but provide an iterator instead as follows:

```
from Bio.PopGen.GenePop.LargeFileParser import read as\
read_large
def count_individuals(fname):
    rec = read_large(open(fname))
```

```
    pop_sizes = []
    for line in rec.data_generator():
        if line == ():
            pop_sizes.append(0)
        else:
            pop_sizes[-1] += 1
    return pop_sizes

print('Individuals per population %s' % ',
'.join([str(len(inds)) for inds in
    count_individuals('hapmap1_auto.gp')])))

print(len(read_large(open('hapmap10.gp')).loci_list))
print(len(read_large(open('hapmap10_auto.gp')).loci_list))
print(len(read_large(open('hapmap10_auto_noofs_ld.gp')).
loci_list))
```

- ❑ The `count_individuals` function shows how you can traverse a Genepop file using the large file parser; while you iterate over it, if you find an empty tuple, it's a marker of a new population. Anything else is an individual, which is composed of tuple (pair) with an individual name and a list of loci (which we will not read here).

- ❑ As stated earlier, individuals per population will return the exact same values.

- ❑ We then print the number of loci on three different files: 10 percent sampling, 10 percent sampling with only autosomes, and 10 percent sampling of autosomes with LD-pruning. The output reflects (which will vary due to stochasticity generating the files) that the first file has more markers than the second file (as the second file is a subset of the first file, removing sex chromosomes and mitochondria) and the last file will have much less markers because it's a LD-pruned subset of the second file.

See also

- ▶ There is actually a Genepop interface on the Web at `http://genepop.curtin.edu.au/` that you can use for manual examples (especially with small files).

Exploring a dataset with Bio.PopGen

In this chapter, we will perform an initial exploratory analysis of one of our generated datasets. We will analyze the 10 percent sampling of chromosome 2 without the offspring. We will look for monomorphic loci (in this case, SNPs) across populations along with how to research minimum allele frequencies and expected heterozygosities.

Getting ready

You will need to have run the previous two recipes and should have the `hapmap10_auto_noofs_2.gp` and `hapmap10_auto_noofs_2.pops` files. We will also use the metadata file downloaded in the first recipe.

For this code to work, you will need to install Genepop from `http://kimura.univ-montp2.fr/~rousset/Genepop.htm`. We will use the interface provided by Biopython to execute Genepop and parse its output files.

There is a notebook with this recipe: `03_PopGen/Exploratory_Analysis.ipynb`, but it will still require running the previous two notebooks in order for the required files to be generated.

How to do it...

Take a look at the following steps:

1. Let's load population names and execute Genepop externally to compute genotypic frequencies as follows:

   ```
   from Bio.PopGen.GenePop import Controller as gpc
   ctrl = gpc.GenePopController()
   my_pops = [l.rstrip() for l in
       open('hapmap10_auto_noofs_2.pops')]
   num_pops = len(my_pops)
   pop_iter, loci_iter =\
   ctrl.calc_allele_genotype_freqs('hapmap10_auto_noofs_2.gp')
   ```

   ❑ First, we will create a controller (an object that allows you to interact with the Genepop application). Then, we will load population names. Finally, we will compute the genotypic information, which may take some time. Our controller will return two iterators, exposing results per population and per loci.

> We will use a relatively small dataset, which makes running Genepop in a single go feasible. If you have a larger dataset, be it in a number of individuals or in the number of loci, you may need to split the file into smaller chunks and run several Genepop instances in parallel, each working with part of the data. We will discuss this in *Chapter 9, Python for Big Genomics Datasets*.

2. Let's go through all loci statistics and retrieve information for each population of fixated alleles, minimum allele frequencies, and number of reads:

```
from collections import defaultdict
fix_pops = [0 for i in range(num_pops)]
num_reads = [defaultdict(int) for i in range(num_pops)]
num_buckets = 20
MAFs = []
for i in range(num_pops):
    MAFs.append([0] * num_buckets)
for locus_data in loci_iter:
    locus_name = locus_data[0]
    allele_list = locus_data[1]
    pop_of_loci = locus_data[2]
    for i in range(num_pops):
        locus_num_reads = pop_of_loci[i][2]
        num_reads[i][locus_num_reads] += 1
        maf = min(pop_of_loci[i][1])
        if maf == 0:
            fix_pops[i] += 1
        else:
            bucket = min([num_buckets - 1, int(maf * 2 *
                num_buckets)])
            MAFs[i][bucket] += 1
```

- We initialize three data structures. One to count the number of monomorphic loci per population, another to count the number of reads per loci and per population, and finally, one to hold the minimum allele frequency per population.

- The minimum allele frequency will be held in a bin-wise fashion (values between 0 and 0.025 will go in the first bin, values between 0.025 and 0.05 will go in the second bin, and so on, until 0.475 and 0.5).

- We then go through the loci iterator provided in the previous entry. We extract the locus name, list of alleles for the loci, and per population information for the locus. We extract the number of alleles read (twice the number of samples as a rule) and allele frequencies. We use the minimum to calculate the MAF and infer whether the locus is monomorphic (MAF of 0).

3. Let's plot the results as follows:

```
import numpy as np
import seaborn as sns
import matplotlib.pyplot as plt
fig, axs = plt.subplots(3, figsize=(16, 9), squeeze=False)
axs[0, 0].bar(range(num_pops), fix_pops)
axs[0, 0].set_xlim(0, 11)
axs[0, 0].set_xticks(0.5 + np.arange(num_pops))
axs[0, 0].set_xticklabels(my_pops)
axs[0, 0].set_title('Monomorphic positions')

axs[1, 0].bar(range(num_pops), [np.max(vals.keys()) for
    vals in num_reads])
axs[1, 0].set_xlim(0, 11)
axs[1, 0].set_xticks(0.5 + np.arange(num_pops))
axs[1, 0].set_xticklabels(my_pops)
axs[1, 0].set_title('Maximum number of allele reads per
loci')

for pop in [0, 7, 8]:
    axs[2, 0].plot(MAFs[pop], label=my_pops[pop])
axs[2, 0].legend()
axs[2, 0].set_xticks(range(num_buckets + 1))
axs[2, 0].set_xticklabels(['%.3f' % (x / (num_buckets *
    2.)) for x in range(num_buckets + 1)])
axs[2, 0].set_title('MAF bundled in bins of 0.025')
```

❑ The reason we import seaborn is because it will change the default look of plots. By conscious decision, the default matplotlib look is not very stylish. Other than this, all code should be fairly easy to interpret.

❑ The output is seen in the following figure and includes three subplots: one with the number of fixated (monomorphic) alleles per population, another with the maximum number of allele reads per loci (mostly, a proxy of a number of individuals processed per population), and finally, the distribution of MAF for three populations. This is the population chosen with the least number of samples (ASW and MEX) and one with the most number of samples (MKK). This was done to illustrate sampling effects; ASW and MEX are bumpy. This is most probably due to less sample size influencing the distribution of MAFs (less values become possible), whereas MKK is smoother:

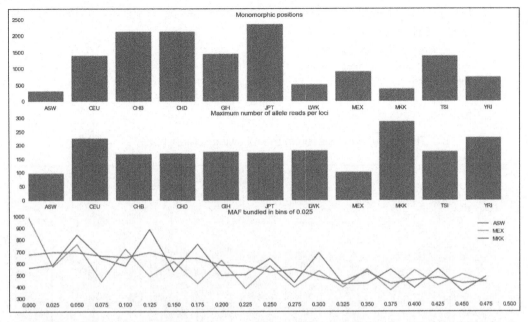

Figure 1: Three subplots: the top one includes a count of fixated SNPs per population, the second one the maximum allele reads per population, and the bottom one the distribution of MAF for three of the eleven populations

4. Let's now traverse the same result, but population-wise to compute the expected heterozygosities per population and per loci:

```
exp_hes = []
for pop_data in pop_iter:
    pop_name, allele = pop_data
    print(pop_name)
    exp_vals = []
    for locus_name, vals in allele.items():
        geno_list, heterozygosity, allele_cnts, summary =\
vals
        cexp_ho, cobs_ho, cexp_he, cobs_he = heterozygosity
        exp_vals.append(cexp_he / (cexp_he + cexp_ho))
    exp_hes.append(exp_vals)
```

- Now, we traverse the iterator with the population information. We extract the population name (remember from the previous recipe that this is not very informative) and then go through each locus and extract the expected number of homozygotes and convert these to the expected heterozygosity.

- Note that this iterator is still somewhat memory intensive; it will load all loci for a single population in memory; this is not very scalable if you have millions of SNPs.

5. Let's plot the distribution of expected heterozygosities per population as follows:

```
fig = plt.figure(figsize=(16, 9))
ax = fig.add_subplot(111)
sns.boxplot(exp_hes, ax=ax)
ax.set_title('Distribution of expected Heterozygosity')
ax.set_xticks(1 + np.arange(num_pops))
ax.set_xticklabels(my_pops)
```

❑ The output can be seen in the following screenshot:

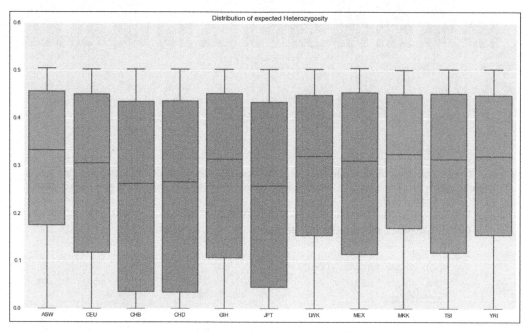

Figure 2: The distribution of expected heterozygosity across eleven populations of the HapMap project for chromosome 2 subsampled at 10 percent

There's more...

The truth is that for population genetic analysis, nothing beats R; you are definitely encouraged to take a look at the existing R libraries for population genetics. Do not forget that there is a Python-R bridge, which is discussed in *Chapter 1, Python and the Surrounding Software Ecology.*

Many of the analysis presented here will be computationally costly if done to bigger datasets (remember that we are only using chromosome 2 subsampled at 10 percent). The final chapter will discuss ways to address this.

See also

▸   There is list of R packages for statistical genetics available at `http://cran.r-project.org/web/views/Genetics.html`

▸   If you need to know more about population genetics, I recommend the book *Principles of Population Genetics, Daniel L. Hartl* and *Andrew G. Clark, Sinauer Associates*

Computing F-statistics

Nearly 100 years ago, Sewall Wright developed F-statistics to quantify inbreeding effects at a certain level of population subdivision. F_{ST} is the most widely used of these statistics and is mostly interpreted as the genetic variation caused by the population structure.

Getting ready

You will need to have run the first two recipes and should have the `hapmap10_auto_noofs_2.gp` and `hapmap10_auto_noofs_2.pops` files. We will also use the metadata file downloaded in the first recipe. For the type of comparison that we will perform here, it's important to assure that there is little relatedness among sampled individuals, so we want to remove the offspring at the very least. For efficiency, we will use only chromosome 2 subsampled at 10 percent.

For this code to work, you will need to install Genepop from `http://kimura.univ-montp2.fr/~rousset/Genepop.htm`. We will use the interface provided by Biopython to execute Genepop and parse its output files. These requirements are the same as the previous recipe.

There is a notebook with the `03_PopGen/F-stats.ipynb` recipe, but it will still require you to run the first two notebooks in order to generate files that are required.

How to do it...

Take a look at the following steps:

1.   First, let's compute F-statistics (F_{ST}, F_{IS}, and F_{IT}) for our dataset with all 11 populations as follows:

```
from Bio.PopGen.GenePop import Controller as gpc
my_pops = [l.rstrip() for l in
    open('hapmap10_auto_noofs_2.pops')]
num_pops = len(my_pops)
ctrl = gpc.GenePopController()
(multi_fis, multi_fst, multi_fit), f_iter =\
ctrl.calc_fst_all('hapmap10_auto_noofs_2.gp')
print(multi_fis, multi_fst, multi_fit)
```

❏ As with the previous recipe, we will first load population names and initialize a controller to interact with the Genepop application. Then, we will calculate various F-statistics.

❏ The function will return a loci iterator that returns several F-statistics per loci as the last parameter. You may be tempted to compute the average F-statistic by looping through the iterator; while this is interesting, the multilocus F-statistics are not trivial to compute. For example, you will want to give more weightage to a loci with a larger MAF. Genepop provides multilocus F_{IS}, F_{ST}, and F_{IT} as its first three parameters before the iterator.

2. Let's traverse loci results and put them on arrays as follows:

```
fst_vals = []
fis_vals = []
fit_vals = []
for f_case in f_iter:
    name, fis, fst, fit, qinter, qintra = f_case
    fst_vals.append(fst)
    fis_vals.append(fis)
    fit_vals.append(fit)
```

❏ This code assumes that you can fit all the values in memory. If your dataset is large, you may want to consider using a slightly more sophisticated approach. Refer to *Chapter 9, Python for Big Genomics Datasets* for ideas.

3. Let's plot the summary distributions:

```
sns.set_style("whitegrid")
fig = plt.figure()
ax = fig.add_subplot(1, 1, 1)
ax.hist(fst_vals, 50, color='r')
ax.set_title('F_ST, F_IS and F_IS distributions')
ax.set_xlabel('F_ST')
ax = fig.add_subplot(2, 2, 2)
sns.violinplot([fis_vals, fit_vals], ax=ax, vert=False)
ax.set_yticklabels(['F_IS', 'F_IT'])
ax.set_xlim(-.1, 0.4)
```

❑ The results can be seen in *Figure 3*, which depicts distributions of F_{ST}, F_{IS}, and F_{IT} across chromosome 2:

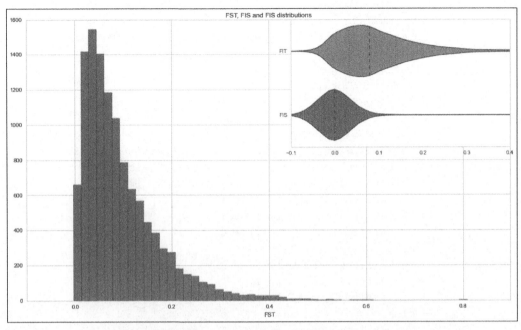

Figure 3: In the large chart, we see a histogram of F_{ST}; the small chart has violin plots for F_{IS} and F_{IT}

4. Let's compute average pair-wise F_{STs} among all populations, as shown in the following code:

```
fpair_iter, avg =\
ctrl.calc_fst_pair('hapmap10_auto_noofs_2.gp')
```

❑ Remember that now we will be comparing all pairs of populations, so we will have per SNP and 55 F_{ST} values (the number of combinations possible with our 11 HapMap populations). You can access all these values on `fpair_iter`. However, for now, we will concentrate on `avg`, which reports the multilocus pair-wise F_{ST} for all 55 combinations.

5. Let's plot the distance matrix across populations based on the multilocus pair-wise F_{ST} as follows:

```python
min_pair = min(avg.values())
max_pair = max(avg.values())
arr = np.ones((num_pops - 1, num_pops - 1, 3), dtype=float)
sns.set_style("white")
fig = plt.figure(figsize=(16, 9))
ax = fig.add_subplot(111)
for row in range(num_pops - 1):
    for col in range(row + 1, num_pops):
        val = avg[(col, row)]
        norm_val = (val - min_pair) / (max_pair - min_pair)
        ax.text(col - 1, row, '%.3f' % val, ha='center')
        if norm_val == 0.0:
            arr[row, col - 1, 0] = 1
            arr[row, col - 1, 1] = 1
            arr[row, col - 1, 2] = 0
        elif norm_val == 1.0:
            arr[row, col - 1, 0] = 1
            arr[row, col - 1, 1] = 0
            arr[row, col - 1, 2] = 1
        else:
            arr[row, col - 1, 0] = 1 - norm_val
            arr[row, col - 1, 1] = 1
            arr[row, col - 1, 2] = 1
ax.imshow(arr, interpolation='none')
ax.set_xticks(range(num_pops - 1))
ax.set_xticklabels(my_pops[1:])
ax.set_yticks(range(num_pops - 1))
ax.set_yticklabels(my_pops[:-1])
```

❑ In *Figure 4,* we will draw an upper triangular matrix, where the background color of a cell represents the measure of differentiation; white means less divergent (lower F_{ST}) and blue means more divergent (higher F_{ST}). The lowest value between CHB and CHD is represented in yellow color and the biggest value between JPT and YRI is represented in magenta color. The value on each cell is the average pair-wise F_{ST} between these two populations:

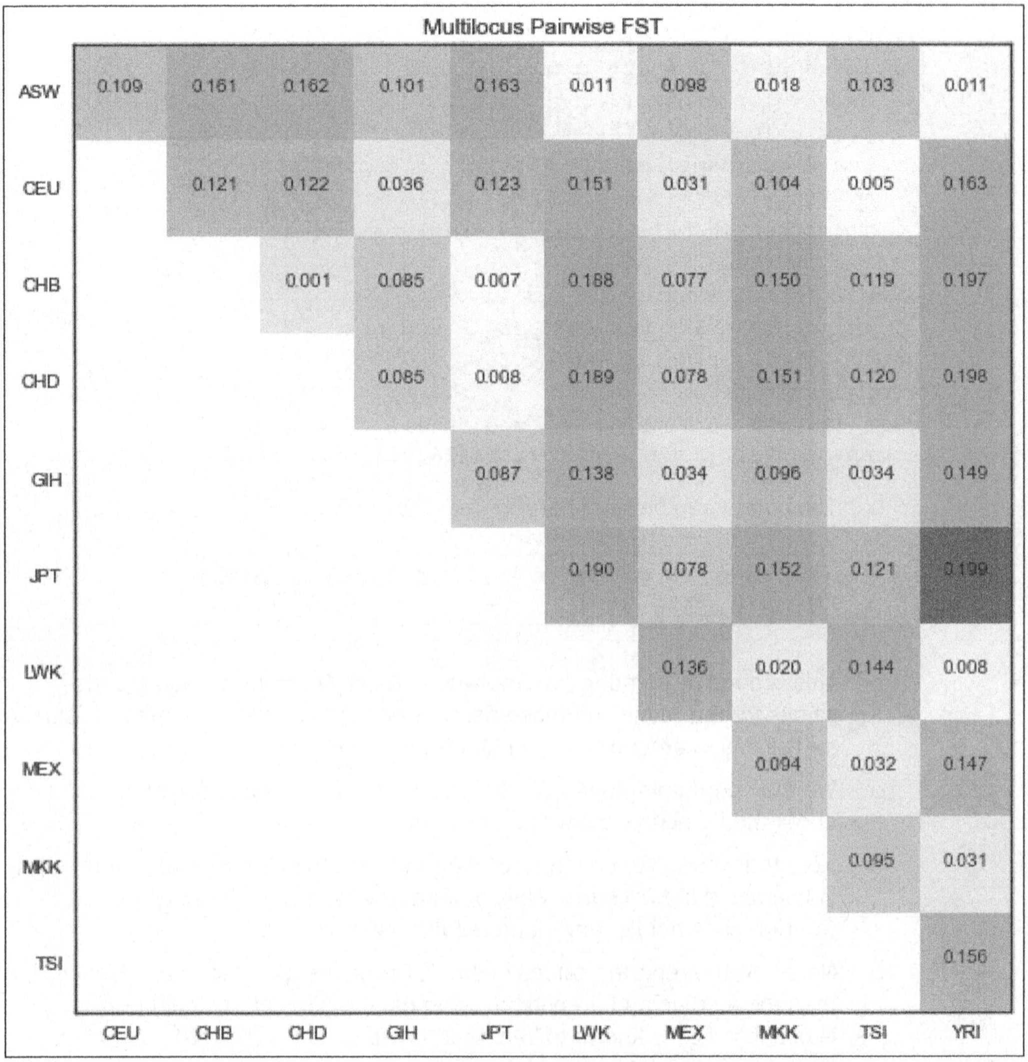

	CEU	CHB	CHD	GIH	JPT	LWK	MEX	MKK	TSI	YRI
ASW	0.109	0.161	0.162	0.101	0.163	0.011	0.098	0.018	0.103	0.011
CEU		0.121	0.122	0.036	0.123	0.151	0.031	0.104	0.005	0.163
CHB			0.001	0.085	0.007	0.188	0.077	0.150	0.119	0.197
CHD				0.085	0.008	0.189	0.078	0.151	0.120	0.198
GIH					0.087	0.138	0.034	0.096	0.034	0.149
JPT						0.190	0.078	0.152	0.121	0.199
LWK							0.136	0.020	0.144	0.008
MEX								0.094	0.032	0.147
MKK									0.095	0.031
TSI										0.156

Multilocus Pairwise FST

Figure 4: The average pair-wise F_{ST} across 11 populations of the HapMap project for chromosome 2

6. Finally, let's check the values of pair-wise F_{ST} comparisons between Yorubans (YRI) and Utah residents with Northwest European ancestry (CEU) around the Lactase (LCT) gene that resides on chromosome 2:

```
pop_ceu = my_pops.index('CEU')
pop_yri = my_pops.index('YRI')
start_pos = 136261886  # b36end_pos = 136350481
all_fsts = []
inside_fsts = []
for locus_pfst in fpair_iter:
    name = locus_pfst[0]
    pfst = locus_pfst[1]
    pos = int(name.split('/')[-1])  # dependent
    my_fst = pfst[(pop_yri, pop_ceu)]
    if my_fst == '-':  # Can be this
        continue
    all_fsts.append(my_fst)
    if pos >= start_pos and pos <= end_pos:
        inside_fsts.append(my_fst)
print(inside_fsts)
print('%.2f/%.2f/%.2f' % (np.median(all_fsts),
    np.mean(all_fsts), np.percentile(all_fsts, 90)))
```

❑ The output can be seen here:

```
[0.1346, 0.6703, 0.7317, 0.1432, 0.2485]
0.08/0.13/0.33
```

❑ This is done by iterating over the whole result. On one side, you get the values for the whole chromosome, whereas on the other, you get the values for the region around LCT (and MCM6, another gene in the neighborhood).

❑ We then print pair-wise F_{STs} in the region and some statistical information about the F_{ST} across the whole chromosome.

❑ Note that you will definitely have different results in the region; your random subsampled file will surely have other markers. If you are unlucky enough, you may even not get any markers (although this is unlikely).

❑ Note how the reported values in the LCT area are generally much higher than the median and even in some cases, the ninetieth percentile. This is because LCT is known to be under the selection of the CEU population (giving that population the ability to digest milk into adult age). F_{ST} is a statistic that can help you perform selection scans (to find genes that may be under selection) and genes that are under directional selection will probably have SNPs with high F_{ST}.

See also

- ► F-statistics is an immensely complex topic and I will direct you firstly to the Wikipedia page at `http://en.wikipedia.org/wiki/F-statistics`.

- ► A very good explanation can be found on Holsinger, Weir paper, and Nature Reviews Genetics (genetics in geographically structured populations: defining, estimating, and interpreting F_{ST}) at `http://www.nature.com/nrg/journal/v10/n9/abs/nrg2611.html`

Performing Principal Components Analysis

Principal Components Analysis (**PCA**) is a statistical procedure to perform a reduction of dimension of a number of variables to a smaller subset that is linearly uncorrelated. Its practical application in population genetics is assisting the visualization of relationships of individuals that is being studied.

While most of the recipes in this chapter make use of Python as a "glue language" (Python calls external applications that actually do most of the work) with PCA, we have an option, that is, we can either use an external application (for example, EIGENSOFT smartpca) or use scikit-learn and perform everything on Python. We will perform both.

Getting ready

You will need to run the first recipe in order to use the `hapmap10_auto_noofs_ld_12` PLINK file (with alleles recoded as 1 and 2). PCA requires LD-pruned markers; we will not risk using the offspring here because it will probably bias the result. We will use the recoded PLINK file with alleles as 1 and 2 because this makes processing easier with smartpca and scikit-learn.

As with the second recipe, if you are not using Docker, you will also be using some of the code that I have produced. You can find this code at `https://github.com/tiagoantao/pygenomics`. You can install it with

pip install pygenomics

For this recipe, you will need to download EIGENSOFT (`http://www.hsph.harvard.edu/alkes-price/software/`), which includes the smartpca application that we will use.

There is a notebook in the `03_PopGen/PCA.ipynb` recipe, but you will still need to run the first recipe.

How to do it...

Take a look at the following steps:

1. Let's load the metadata as follows:

```
f = open('relationships_w_pops_121708.txt')
ind_pop = {}
f.readline()   # header
for l in f:
    toks = l.rstrip().split('\t')
    fam_id = toks[0]
    ind_id = toks[1]
    pop = toks[-1]
    ind_pop['/'.join([fam_id, ind_id])] = pop
f.close()
ind_pop['2469/NA20281'] = ind_pop['2805/NA20281']
```

> ❏ In this case, we will add an entry that is consistent with what is available on the PINK file.

2. Let's convert the PLINK file to the EIGENSOFT format:

```
from genomics.popgen.plink.convert import to_eigen
to_eigen('hapmap10_auto_noofs_ld_12',
    'hapmap10_auto_noofs_ld_12')
```

> ❏ This uses a function that I have written to convert from PLINK to the EIGENSOFT format. This is mostly text manipulation, not precisely the most exciting code.

3. Now, we will run smartpca and parse its results as follows:

```
from genomics.popgen.pca import smart
ctrl = smart.SmartPCAController('hapmap10_auto_noofs_ld_12')
ctrl.run()
wei, wei_perc, ind_comp =\
smart.parse_evec('hapmap10_auto_noofs_ld_12.evec',
    'hapmap10_auto_noofs_ld_12.eval')
```

> ❏ Again, this will use a couple of functions from pygenomics to control smartpca and then to parse the output. The code is typical for this kind of operations, and while you are invited to inspect it, be aware that it's quite straightforward.

> ❏ The parse function will return PCA weights (which we will not use, but you should inspect), normalized weights, and then principal components (usually up to PC 10) per individual.

4. Then, we plot PC1 and PC2, as shown in the following code:

```
from genomics.popgen.pca import plot
plot.render_pca(ind_comp, 1, 2, cluster=ind_pop)
```

- ❑ This will produce the following figure. We will supply the plotting function and the population information retrieved from the metadata, which allows you to plot each population with a different color.

- ❑ The results are very similar to published results; we will find four groups. Most Asian populations are located on top, the African populations are located on the right-hand side, and the European populations are located at the bottom. Two more admixed populations (GIH and MEX) are located in the middle:

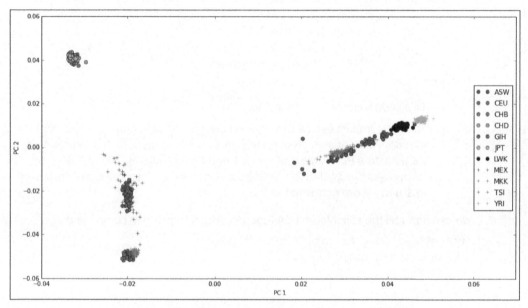

Figure 5: PC1 and PC2 of the HapMap data as produced by smartpca

5. Now, let's turn to a PCA plot produced by Python libraries only. To be able to run scikit-learn PCA on our data, let's get the individual order on the PED file and the number of SNPs first as follows:

```
f = open('hapmap10_auto_noofs_ld_12.ped')
ninds = 0
ind_order = []
for l in f:
    ninds += 1
    toks = l[:100].replace(' ', '\t').split('\t')
    fam_id = toks[0]
```

```
        ind_id = toks[1]
        ind_order.append('%s/%s' % (fam_id, ind_id))
    nsnps = (len(line.replace(' ', '\t').split('\t')) - 6) // 2
    f.close()
```

6. Then, we create an array required for the PCA function reading in the PED file:

```
import numpy as np
pca_array = np.empty((ninds, nsnps), dtype=int)

f = open('hapmap10_auto_noofs_ld_12.ped')
for ind, l in enumerate(f):
    snps = l.replace(' ', '\t').split('\t')[6:]
    for pos in range(len(snps) // 2):
        a1 = int(snps[2 * pos])
        a2 = int(snps[2 * pos])
        my_code = a1 + a2 - 2
        pca_array[ind, pos] = my_code
f.close()
```

- ❏ This code will be slow to execute.

- ❏ The most import part of the code is coding of alleles, and the ability for PCA to produce meaningful results rely on a good coding here. Remember that we will use a PLINK file that has a 1 and 2 allele coding. We will use the following strategy, that is 11s are converted to 0, 12 (and 21) are converted to 1 and 22 are converted to 2.

7. We can now call the scikit-learn PCA function, which requests 8 components:

```
from sklearn.decomposition import PCA
my_pca = PCA(n_components=8)
my_pca.fit(pca_array)
trans = my_pca.transform(pca_array)
```

8. Finally, let's print eight PCs as follows:

```
sc_ind_comp = {}
for i, ind_pca in enumerate(trans):
    sc_ind_comp[ind_order[i]] = ind_pca
plot.render_pca_eight(sc_ind_comp, cluster=ind_pop)
```

- ❏ We will use a different function to perform the plotting here; you will able to see up to component 8.

❑ The result is qualitatively similar to the smartpca version (it will be a worrying situation if it had been otherwise). Note that this is a mirror image from the previous figure; swapping signals on PCA is not a major issue at all:

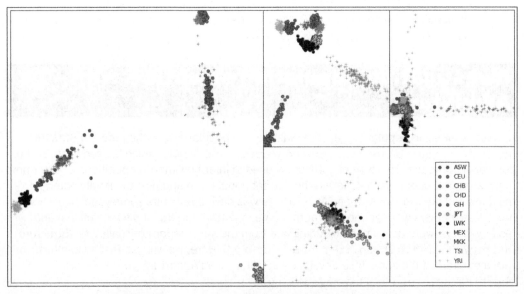

Figure 6: PC1 to PC8 of the HapMap data as produced by scikit-learn

There's more...

An interesting question here is which method should you use? smartpca or scikit-learn? The results are similar, so if you are performing your own analysis, you are free to choose. However, if you publish your results in a scientific journal, smartpca is probably a safer choice because it's based on the published piece of software in the field of genetics; reviewers will probably prefer this.

See also

▸ The paper that probably popularized the use of PCA in genetics was Novembre et al *Genes mirror geography within Europe* on Nature, where a PCA of Europeans map almost perfectly to the map of Europe at `http://www.nature.com/nature/journal/v456/n7218/abs/nature07331.html`. Note that there is nothing in PCA that assures it will map to geographical features (just check our PCA earlier).

▸ The smartpca is described in Patterson et al, *Population structure and eigenanalysis, PLoS Genetics* at `http://journals.plos.org/plosgenetics/article?id=10.1371/journal.pgen.0020190`.

- ▶ A discussion of the meaning of PCA can be found in McVean's paper on, *A Genealogical Interpretation of Principal Components Analysis, PLoS Genetics* at `http://journals.plos.org/plosgenetics/article?id=10.1371/journal.pgen.1000686`.

- ▶ As usual, the Wikipedia page is nicely written at `http://en.wikipedia.org/wiki/Principal_component_analysis`.

Investigating population structure with Admixture

A typical analysis in population genetics was the one popularized by the program Structure (`http://pritchardlab.stanford.edu/structure.html`), which is used to study the population structure. This type of software is used to infer how many populations exist (or how many ancestral populations generate the current population) and how to identify potential migrants and admixed individuals. As Structure was developed quite some years ago when much less markers were genotyped (at that time, mostly a handful of microsatellites) and faster versions were developed, including one from the same laboratory called *fastStructure* (`http://rajanil.github.io/fastStructure/`). Here, we will use Python to interface with a program of the same type developed at UCLA called Admixture (`https://www.genetics.ucla.edu/software/admixture/`).

Getting ready

You will need to run the first recipe in order to use the `hapmap10_auto_noofs_ld` binary PLINK file. Again, we will use a 10 percent subsampling of autosomes LD-pruned with no offspring.

As in the second recipe, if you are not using Docker, you will have to download the code that I have produced; you can find these code files at `https://github.com/tiagoantao/pygenomics`. You can install it with

```
pip install pygenomics
```

In theory, for this recipe, you will need to download Admixture (`https://www.genetics.ucla.edu/software/admixture/`). However, in this case, I do provide the outputs of running Admixture on the HapMap data that we will use because running Admixture takes a lot of time. You can either use the results available or run Admixture yourself.

There is a notebook in the `03_PopGen/Admixture.ipynb` recipe, but you will still need to run the first recipe.

How to do it...

Take a look at the following steps:

1. First, let's define our K (a number of ancestral populations) range of interest as follows:

```
k_range = range(2, 10)   # 2..9
```

2. Let's run Admixture for all our Ks (alternatively, you can skip this step and use the example data provided):

```
for k in k_range:
    os.system('admixture --cv=10 hapmap10_auto_noofs_ld.bed %d > admix.%d' % (k, k))
```

This is the worst possible way of running Admixture and will probably take more than 3 hours if you do it like this because it will run all Ks from two to nine in a sequence. There are two things that you can do to speed this up: use the multithreaded option (-j), which Admixture provides or run several applications in parallel. Here, I have to assume a worst case scenario where you only have a single core and thread available, but you should be able to run this more efficiently by parallelizing. We will discuss this issue at length in the last chapter.

3. We will need the order of individuals in the PLINK file, as Admixture outputs individual results in this order:

```
f = open('hapmap10_auto_noofs_ld.fam')
ind_order = []
for l in f:
    toks = l.rstrip().replace(' ', '\t').split('\t')
    fam_id = toks[0]
    ind_id = toks[1]
    ind_order.append((fam_id, ind_id))
f.close()
```

4. The cross-validation error gives a measure of the "best" K as follows:

```
import matplotlib.pyplot as plt
CVs = []
for k in k_range:
    f = open('admix.%d' % k)
    for l in f:
        if l.find('CV error') > -1:
            CVs.append(float(l.rstrip().split(' ')[-1]))
            break
    f.close()
```

```
fig = plt.figure(figsize=(16, 9))
ax = fig.add_subplot(111)
ax.set_title('Cross-Validation error')
ax.set_xlabel('K')
ax.plot(k_range, CVs)
```

❑ The following figure plots the CV between a K of 2 and 9; lower is better. It should be clear from this figure that we should run maybe some more Ks (indeed, we have 11 populations; if not more, we should at least run up to 11), but due to computation costs, we stopped at 9.

❑ It will be a very technical debate on whether there is a thing as the "best" K, but modern scientific literature suggests that there may not be something as a "best" K; these results are worthy of some interpretation. I think it's important that you are aware of this before you go ahead and interpret K results:

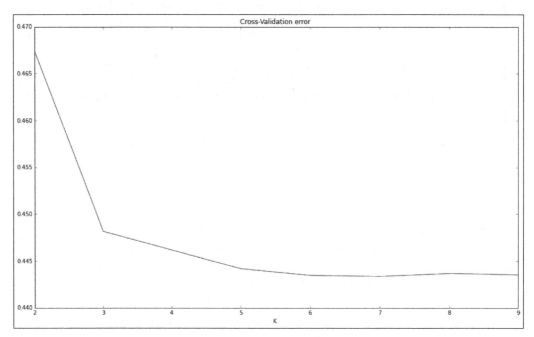

Figure 7: The error per K

5. We will need the metadata for the population information:

```
f = open('relationships_w_pops_121708.txt')
pop_ind = defaultdict(list)
f.readline()   # header
for l in f:
    toks = l.rstrip().split('\t')
    fam_id = toks[0]
```

```
            ind_id = toks[1]
            if (fam_id, ind_id) not in ind_order:
                continue
            mom = toks[2]
            dad = toks[3]
            if mom != '0' or dad != '0':
                continue
            pop = toks[-1]
            pop_ind[pop].append((fam_id, ind_id))
    f.close()
```

- ❏ We ignore individuals that are not in the PLINK file.

6. Let's load the individual component as follows:

```
def load_Q(fname, ind_order):
    ind_comps = {}
    f = open(fname)
    for i, l in enumerate(f):
        comps = [float(x) for x in l.rstrip().split(' ')]
        ind_comps[ind_order[i]] = comps
    f.close()
    return ind_comps
comps = {}
for k in k_range:
    comps[k] = load_Q('hapmap10_auto_noofs_ld.%d.Q' % k,
ind_order)
```

- ❏ Admixture produces a file with the ancestral component per individual (for an example, see any of the generated Q files); there will be as many components as the number of Ks that you decided to study.

- ❏ Here, we will load the Q file for all Ks that we studied and store them in a dictionary where the individual ID is the key.

7. Then, we cluster individuals as follows:

```
from genomics.popgen.admix import cluster
ordering = {}
for k in k_range:
    ordering[k] = cluster(comps[k], pop_ind)
```

- ❏ Remember that individuals were given components of ancestral populations by Admixture; we would like to order them as per their similarity in terms of ancestral components (not by their order in the PLINK file). This is not a trivial exercise and does require any clustering algorithm.

- ❏ Furthermore, we do not want to order all of them; we want to order them in each population and then order each population accordingly.

- ❑ For this purpose, I have some clustering code available at `https://github.com/tiagoantao/pygenomics/blob/master/genomics/popgen/admix/__init__.py`. This is far from perfect, but allows you to perform some plotting that still looks reasonable. My code makes use of the SciPy clustering code. In this case, I suggest you to take a look (by the way, it's not very difficult to improve on it).

8. With a sensible individual order, we can now plot the Admixture:

```
from genomics.popgen.admix import plot
plot.single(comps[4], ordering[4])
fig = plt.figure(figsize=(16, 9))
plot.stacked(comps, ordering[7], fig)
```

- ❑ This will produce two charts; the second chart is seen in the following figure (the first figure is actually a variation of the third Admixture plot from the top).

- ❑ The first figure of K=4 requires components per individual and its order. It will plot all individuals ordered and split by population.

- ❑ The second figure will perform a set of stacked plots of Admixture from K 2 to 9. It does require a figure object (as the dimension of this figure can vary widely with the number of stacked Admixtures that you require). The individual order will typically follow one of the Ks (we have chosen K of 7 here):

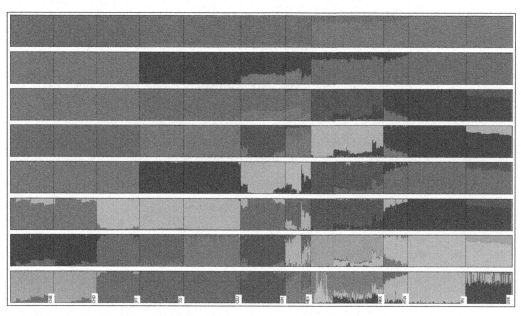

Figure 8: Stacked Admixture plot (between K of 2 and 9) for the HapMap example

There's more...

Unfortunately, you cannot run a single instance of Admixture to get a result. The best practice is to actually run 100 instances and get the one with the best log likelihood (which is reported in the Admixture output). Obviously, I cannot ask you to run 100 instances times 7 different Ks for this recipe (we are talking about two weeks of computation), but you will probably have to perform this if you want to have publishable results. A cluster (or at least a very good machine) is required to run this. Obviously, you can use Python to go through outputs and select the best log likelihood. After selecting the result with the best log likelihood for each K, you can easily apply this recipe to plot the output.

5
Population Genetics Simulation

In this chapter, we will cover the following recipes:

- ▶ Introducing forward-time simulations
- ▶ Simulating selection
- ▶ Simulating population structure using island and stepping-stone models
- ▶ Modeling complex demographic scenarios
- ▶ Simulating the coalescent with Biopython and fastsimcoal

Introduction

In the previous chapter, we used Python to analyze population genetics datasets based on real data. In this chapter, we will see how to use Python to simulate population genetics data. From teaching to developing new statistical methods or to analyze the performance of existing methods, simulated datasets have plenty of applications.

There are two kinds of simulation. One is coalescent simulation that goes backwards in time. Second is forward time. As the name implies, it simulates going forward. The coalescent simulation is computationally less expensive because only the most recent generation of individuals need to be completely rendered; previous generations only need parents of the previous generation to be maintained. On the other hand, this severely limits what can be simulated because we need to complete populations to make decisions on e.g. which individuals mate. Forward time simulations are computationally more demanding and normally more complex to code, but they allow you to have much more flexibility.

In this chapter, we will use the Python-based, forward-time simulator called simuPOP to model very complex scenarios and see how we can analyze its results. We will also have one recipe on the coalescent simulation. Be aware that you will need to know about basic population genetics to understand this chapter.

Introducing forward-time simulations

We will start with a simple recipe to code the bare minimum with simuPOP. simuPOP is probably the most flexible and powerful forward-time simulator available and is Python-based. You will be able to simulate almost anything in terms of demography and genomics, save for complex genome structural variation (for example, inversions or translocations).

Getting ready

simuPOP may apparently be difficult, but it will make sense if you understand its event-oriented model. As you would expect, there is a meta population composed of individuals with a predefined genomic structure. Starting with an initial population that you prepare, a set of initial operators is applied. Then every time a generation ticks, a set of pre-operators are applied, followed by a mating step that generates the new population for the next cycle. This is followed by a final set of postoperators that are applied again. This cycle (preoperations, mating, and postoperations) repeats for as many generations that you desire.

The most important part of getting ready is preparing yourself for the preceding model; we will go through a simple example now. As usual, you can find this on the `04_PopSim/Basic_SimuPOP.ipynb` notebook.

How to do it...

Take a look at the following steps:

1. Let's start by initializing variables and basic data structures as follows:

```
from collections import OrderedDict
num_loci = 10
pop_size = 100
num_gens = 10
init_ops = OrderedDict()
pre_ops = OrderedDict()
post_ops = OrderedDict()
```

   - So, we specify that we want to simulate 10 loci, a population size of 100, and just 10 generations. We then prepare three ordered dictionaries. These will maintain our operators.

- ❏ Note that we use `OrderedDict()`. Ordered dictionaries will return keys in the order that they were inserted. This is important because the order of operators is relevant, for example, if we print the result of a statistic, this will be dependent on it being computed before.

2. We will now create a population object and basic operators as follows:

```
import simuPOP as sp
pops = sp.Population(pop_size, loci=[1] * num_loci)
init_ops['Sex'] = sp.InitSex()
init_ops['Freq'] = sp.InitGenotype(freq=[0.5, 0.5])
post_ops['Stat-freq'] = sp.Stat(alleleFreq=sp.ALL_AVAIL)
post_ops['Stat-freq-eval'] = sp.PyEval(r"'%d %.2f\n' % (gen,
    alleleFreq[0][0])")
mating_scheme = sp.RandomMating()
```

- ❏ We start by creating the meta population with a single deme of the required size and 10 independent loci.

- ❏ Then, create two operators to initialize the population; one operator initializes the sex of all individuals (the default is two sexes with a probability of 50 percent to be assigned to either male or female). The other operator initializes all the loci with two alleles (maybe of an SNP) with a frequency of 50 percent for each allele.

- ❏ We also create two postmating operators: one to calculate allele frequencies for all loci and another to just print the allele frequency of loci 0 and allele 0. These will be executed in all generations.

- ❏ Finally, we specify the standard random mating. This is a very basic model.

3. Let's run the simulator for our basic scenario with a single replicate:

```
sim = sp.Simulator(pops, rep=1)
sim.evolve(initOps=init_ops.values(),
    preOps=pre_ops.values(), postOps=post_ops.values(),
    matingScheme=mating_scheme, gen=num_gens)
```

- ❏ We will create a simulator object that will be responsible for evolving our population. We will specify that we just want a single replicate. Having many replicates is a more common situation, which we will address in future recipes.

The output that I got is as follows. Note that we are dealing with stochastic process, so you will get different results. This is especially serious if the population size is small. Indeed, one of the most important results in population genetics is that in small populations, the stochastic drift is a very strong factor. All results in this chapter are stochastic in nature, so expect to see different results from the ones presented throughout this chapter. If you want deterministic results, you can use a predetermined random seed, but do not let this trick you into thinking that these are deterministic processes in nature.

```
0 0.46
1 0.43
2 0.41
3 0.43
4 0.42
5 0.47
6 0.51
7 0.52
8 0.54
9 0.44
```

4. Let's perform a simple population genetic analysis using a new simulation model, researching the impact of population size on the loss of heterozygosity over time. We start by developing a mini framework to store and easily access variables of interest:

```
from copy import deepcopy
def init_accumulators(pop, param):
    accumulators = param
    for accumulator in accumulators:
        pop.vars()[accumulator] = []
    return True
def update_accumulator(pop, param):
    accumulator, var = param
    pop.vars()[accumulator].append(deepcopy(pop.vars()[var]))
    return True
```

- This code is the complex part of this recipe; it comprises of two operators. One is to be used as an initialization operator, which will add a variable to the population (as simuPOP allows you to maintain extra variables at the population level). Another function will append results in a pre-operator or post-operator to the variable. This may seem abstract now, but we will make it clear soon.

- We will use `deepcopy` to make sure that we have our own copy of the variable because it can be changed by a future operator.

5. We will need to compute the expected heterozygosity from allelic frequency as follows:

```
def calc_exp_he(pop):
    #assuming bi-allelic markers coded as 0 and 1
    pop.dvars().expHe = {}
    for locus, freqs in pop.dvars().alleleFreq.items():
        f0 = freqs[0]
        pop.dvars().expHe[locus] = 1 - f0**2 - (1 - f0)**2
    return True
init_ops['accumulators'] = sp.PyOperator(init_accumulators,
    param=['num_males', 'exp_he'])
post_ops['Stat-males'] = sp.Stat(numOfMales=True)
post_ops['ExpHe'] = sp.PyOperator(calc_exp_he)
post_ops['male_accumulation'] =
sp.PyOperator(update_accumulator, param=('num_males',
    'numOfMales'))
post_ops['expHe_accumulation'] =
sp.PyOperator(update_accumulator, param=('exp_he',
    'expHe'))
del post_ops['Stat-freq-eval']
```

- Note that we are still using operators specified in the previous execution to initialize sex, genotype, and so on; we will just be adding our ordered dictionaries to them.

- Firstly, we will develop a function to compute our expected heterozygosity from allele frequency for all available loci.

- Then, we will add two accumulators (num_males and exp_he) using an initialization operator.

- We will then add four post-operators (this will be applied after reproduction). One will compute the number of males, another will compute the expected heterozygosity, and the third post-operator will transfer the computation from each generation to a variable that stores the result over time. So, numOfMales is the result of a simuPOP operator that computes the number of males for the current generation, whereas num_males maintains a list of all numOfMales across the whole execution. numOfMales is recomputed and lost on each sim.evolve step, whereas num_of_males is appended.

- The previous strategy cannot be used to store very large variables over the whole simulation, but it works for small variables.

- Note that the expected heterozygosity operator depends on the existing operator to compute allele frequencies that already exists in the dictionary. This has to be executed before the one computing expected heterozygosity (the ordered dictionary assures this).

6. We will now compare two populations with population sizes of 40 and 500:

```
num_gens = 100
pops_500 = sp.Population(500, loci=[1] * num_loci)
sim = sp.Simulator(pops_500, rep=1)
sim.evolve(initOps=init_ops.values(),
    preOps=pre_ops.values(), postOps=post_ops.values(),
    matingScheme=mating_scheme, gen=num_gens)
pop_500_after = deepcopy(sim.population(0))
pops_40 = sp.Population(40, loci=[1] * num_loci)
sim = sp.Simulator(pops_40, rep=1)
sim.evolve(initOps=init_ops.values(),
    preOps=pre_ops.values(), postOps=post_ops.values(),
    matingScheme=mating_scheme, gen=num_gens)
pop_40_after = deepcopy(sim.population(0))
```

7. Let's plot the loss of heterozygosity and the distribution of number of males:

```
import numpy as np
import seaborn as sns
import matplotlib.pyplot as plt
def calc_loci_stat(var, fun):
    stat = []
    for gen_data in var:
        stat.append(fun(gen_data.values()))
    return stat
sns.set_style('white')
fig, axs = plt.subplots(1, 2, figsize=(16, 9), sharey=True,
    squeeze=False)
def plot_pop(ax1, pop):
    for locus in range(num_loci):
        ax1.plot([x[locus] for x in pop.dvars().exp_he],
            color=(0.75, 0.75, 0.75))
    mean_exp_he = calc_loci_stat(pop.dvars().exp_he,
        np.mean)
    ax1.plot(mean_exp_he, color='r')
plot_pop(axs[0, 0], pop_40_after)
plot_pop(axs[0, 1], pop_500_after)
ax = fig.add_subplot(4, 4, 13)
ax.boxplot(pop_40_after.dvars().num_males)
ax = fig.add_subplot(4, 4, 16)
ax.boxplot(pop_500_after.dvars().num_males)
fig.tight_layout()
```

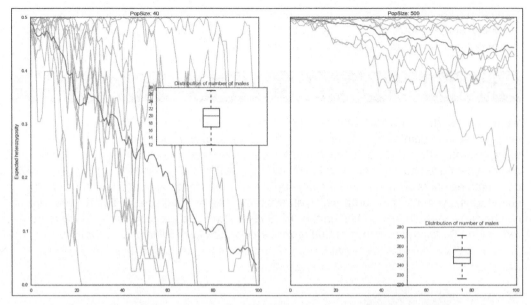

Figure 1: The decline in heterozygosity over time in a population of 40 (left) and 500 (right);
the gray lines are individual markers, whereas the red lines show the mean; the box plots
represent the distribution of number of males in both scenarios

- Major plots show the decrease in expected heterozygosity. They behave exactly as expected: bigger loss in the smaller population with bigger variance. Gray lines depict the individual loci, whereas the red line shows the mean. Note how easy it's to extract the heterozygosity from the preceding code (this is the advantage of using the accumulator framework).

- Note the box plots; they show the distribution of the number of males produced in each generation. Remember that each individual has a probability of 50 percent of being a male, so the number of males will vary in each generation. If you have a very small population, there is a possibility that no males or females are generated in a single cycle. In such cases, the simulator will raise an exception and stop. Try performing a simulation with just 20 individuals and this will eventually happen.

There's more...

simuPOP is a very powerful simulator. Although, we will address some of its features in the next recipes, it's impossible to go through all of them. If you want to simulate linked loci, non-autosomal chromosomes, complex demographies, different sex ratios and models, or mutation, simuPOP will accommodate you. Do not forget to check its website `http://simupop.sourceforge.net/` and to check its great documentation. The user and reference manuals are fantastic. For example, in the documentation, you will find widely used demographic models such as the cosi model of human demographies.

I maintain a set of notebooks to teach population genetics concepts using simuPOP with a lot of code examples. You can find them at `https://github.com/tiagoantao/genomics-notebooks/blob/master/Welcome.ipynb`.

Simulating selection

We will now perform an example of simulating selection with simuPOP. We will perform a simple case with dominant mutation on a single loci and also a complex case with two loci using epistatic effects. The epistatic effect will have a mutation on an SNP. This is required to confer advantage and another mutation on another SNP that adds up to the previous one (but does nothing on its own); this was inspired by the very real case of malaria resistance to the sulfadoxine/pyrimethamine drug, which always requires a mutation on the DHFR gene, which can be enhanced with a mutation on the DHPS gene; if you are interested in knowing more, refer to the *Origin and Evolution of Sulfadoxine Resistant Plasmodium falciparum* article from Vinayak et al on PLoS Pathogens at `http://journals.plos.org/plospathogens/article?id=10.1371/journal.ppat.1000830`.

Getting ready

Read the previous recipe (*Introducing forward-time simulations*) as it will introduce the basic programming framework. If you are using notebooks, this content is in `04_PopSim/Selection.ipynb`.

How to do it...

Take a look at the following steps:

1. Let's start with the boilerplate variable initialization:

```python
from collections import OrderedDict
from copy import deepcopy
import simuPOP as sp
num_loci = 10
pop_size = 1000
num_gens = 101
init_ops = OrderedDict()
pre_ops = OrderedDict()
post_ops = OrderedDict()
def init_accumulators(pop, param):
    accumulators = param
    for accumulator in accumulators:
        pop.vars()[accumulator] = []
    return True
def update_accumulator(pop, param):
```

```
        accumulator, var = param
        pop.vars()[accumulator].append(deepcopy(pop.vars()[var]))
        return True
    pops = sp.Population(pop_size, loci=[1] * num_loci,
        infoFields=['fitness'])
```

- ❑ Most of this code was explained in the previous recipe, but look at the very last line. There is now an `infoFields` parameter in the population. While you can have population variables, you can also have variables for each individual; these are limited to floating point numbers, which are specified in the `infoFields` parameter.

- ❑ We will simulate a fairly big population (1000) to avoid drift effects overpowering selection.

2. To ensure that we can control the number of selected alleles at initialization, let's create a function to perform it:

```
    def create_derived_by_count(pop, param):
        locus, cnt = param
        for i, ind in enumerate(pop.individuals()):
            for marker in range(pop.totNumLoci()):
                if i < cnt and locus == marker:
                    ind.setAllele(1, marker, 0)
                else:
                    ind.setAllele(0, marker, 0)
                ind.setAllele(0, marker, 1)
        return True
```

- ❑ We will use this to initialize our loci under selection (instead of `InitGenotype`, which we will still use for neutral markers).

3. Let's add all operators except for selection:

```
    init_ops['Sex'] = sp.InitSex()
    init_ops['Freq-sel'] =
    sp.PyOperator(create_derived_by_count, param=(0, 10))
    init_ops['Freq-neutral'] = sp.InitGenotype(freq=[0.5, 0.5],
        loci=range(1, num_loci))
    post_ops['Stat-freq'] = sp.Stat(alleleFreq=sp.ALL_AVAIL)
    post_ops['Stat-freq-eval'] = sp.PyEval(r"'%d %.3f\n' %
        (gen, alleleFreq[0][1])", reps=[0], step=10)
    mating_scheme = sp.RandomMating()
```

- ❑ There are selected and neutral loci in this simulation. We will not be using the neutral loci anymore, but it's quite common to simulate a handful selected loci among many neutral loci in order to compare the behavior.

4. Let's add the code for dominant selection at a single locus, store the frequency of the derived (selected) allele, and run the simulation using several replicates:

```
ms = sp.MapSelector(loci=0, fitness={
                (0, 0): 0.90,
                (0, 1): 1,
                (1, 1): 1})
pre_ops['Selection'] = ms
def get_freq_deriv(pop, param):
    marker, name = param
    expHe = {}
    pop.vars()[name] = pop.dvars().alleleFreq[marker][1]
    return True
init_ops['accumulators'] = sp.PyOperator(init_accumulators,
    param=['freq_sel'])
post_ops['FreqSel'] = sp.PyOperator(get_freq_deriv,
    param=(0, 'freqDeriv'))
post_ops['freq_sel_accumulation'] = \
    sp.PyOperator(update_accumulator, param=('freq_sel',
        'freqDeriv'))
sim = sp.Simulator(pops, rep=100)
sim.evolve(initOps=init_ops.values(),
                preOps=pre_ops.values(),
                postOps=post_ops.values(),
                matingScheme=mating_scheme, gen=num_gens)
```

- ❑ We create a fitness operator that will compute the fitness parameter for each individual with a dominant encoding. If you have the selected (coded with 1) mutation, you are fitter than if you are homozygous for the 0 allele. Note that it's quite easy to model a recessive mutation (only benefits the derived homozyguous) or even a heterozygote advantage (only benefits heterozyguous individuals).

- ❑ In this case, we will run 100 replicates (as specified on the simulator initialization). So, there will be 100 independent runs with independent results.

- ❑ For each run, we will store the frequency of the derived (selected) allele. This is the purpose of `get_freq_deriv` and related operators.

5. We can now plot the change in frequency of derived alleles in all 100 independent replicates as follows:

```
import seaborn as sns
import matplotlib.pyplot as plt
sns.set_style('white')
fig = plt.figure(figsize=(16, 9))
```

```
ax = fig.add_subplot(111)
ax.set_title('Frequency of selected alleles in 100
replicates over time')
ax.set_xlabel('Generation')
ax.set_ylabel('Frequency of selected allele')
for pop in sim.populations():
    ax.plot(pop.vars()['freq_sel'])
```

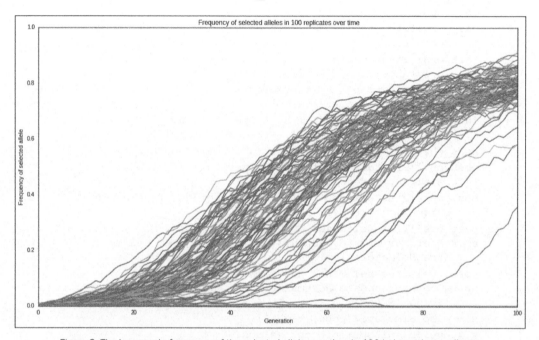

Figure 2: The increase in frequency of the selected allele over time in 100 independent replicates

> ❑ Note that each line represents a trajectory in different replicates. This was
> performed with a large population size; if you try this with a smaller value
> (say 50), you will quite a different pattern due to drift.

6. Let's now look at a complex example involving epistasis between two loci under selection. Here, we will perform 15 replicates, as shown in the following code:

```
pop_size = 5000
num_gens = 100
pops = sp.Population(pop_size, loci=[1] * num_loci,
    infoFields=['fitness'])
def example_epistasis(geno):
    if geno[0] + geno[1] == 0:
        return 0.7
```

```
        elif geno[2] + geno[3] == 0:
            return 0.8
        else:
            return 0.9 + 0.1 * (geno[2] + geno[3] - 1)
init_ops = OrderedDict()
pre_ops = OrderedDict()
post_ops = OrderedDict()
init_ops['Sex'] = sp.InitSex()
init_ops['Freq-sel'] = sp.InitGenotype(freq=[0.99, 0.01],
    loci=[0, 1])
init_ops['Freq-neutral'] = sp.InitGenotype(freq=[0.5, 0.5],
    loci=range(2, num_loci))
pre_ops['Selection'] = sp.PySelector(loci=[0, 1],
    func=example_epistasis)
init_ops['accumulators'] = sp.PyOperator(init_accumulators,
    param=['freq_sel_major', 'freq_sel_minor'])
post_ops['Stat-freq'] = sp.Stat(alleleFreq=sp.ALL_AVAIL)
post_ops['FreqSelMajor'] = sp.PyOperator(get_freq_deriv,
    param=(0, 'FreqSelMajor'))
post_ops['FreqSelMinor'] = sp.PyOperator(get_freq_deriv,
    param=(1, 'FreqSelMinor'))
post_ops['freq_sel_major_accumulation'] =
sp.PyOperator(update_accumulator, param=('freq_sel_major',
    'FreqSelMajor'))
post_ops['freq_sel_minor_accumulation'] = \
sp.PyOperator(update_accumulator, param=('freq_sel_minor',
    'FreqSelMinor'))
sim = sp.Simulator(pops, rep=15)
sim.evolve(initOps=init_ops.values(),
    preOps=pre_ops.values(), postOps=post_ops.values(),
    matingScheme=mating_scheme, gen=num_gens)
```

- ❏ The `example_epistasis` function will take the first two loci to compute the fitness of the individual: 0.7 if it does not have the derived allele at the main loci, 0.8 if it has the derived allele at the main loci, but no derived allele at the secondary loci, 0.9 if it has one derived secondary loci (adding to the main loci), and 1 if it's homozyguous for the derived allele in the secondary loci (again adding to the main one). So, having a mutation just on the secondary is irrelevant (it will stay at 0.7). The main loci is dominant, but the secondary loci is coded as additive (it's better to have two alleles that are derived than just one).

- ❏ We initialize the selected loci separately with a derived frequency of 1 percent, whereas the neutral starts at 50 percent.

- ❏ We also track both our selected alleles using the usual framework.

7. Let's plot the frequencies of both selected alleles (the main and the secondary loci):

```
fig = plt.figure(figsize=(16, 9))
ax1 = fig.add_subplot(111)
ax.set_xlabel('Generation')
ax.set_ylabel('Frequency of selected allele')
ax1.set_title('Frequency of selected alleles (principal and
supporting) over time in 15 replicates')
for pop in sim.populations():

    ax1.plot(pop.vars()['freq_sel_major'])
    ax1.plot(pop.vars()['freq_sel_minor'], '-')
```

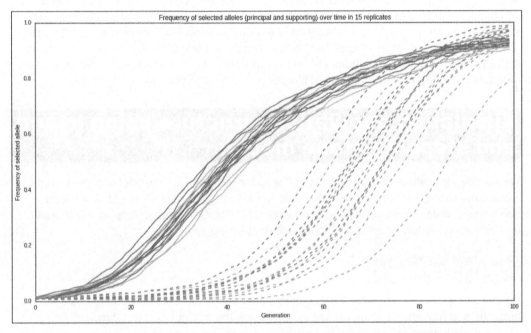

Figure 3: Frequency of selected alleles (principal and supporting loci) over time in 15 replicates; the main allele is coded with a straight line and the secondary allele as a dashed line

❑ In the preceding figure, you can see the dynamic of both alleles: the main allele (shown as a straight line) and the secondary allele (shown as a dashed line).

There's more...

If you run multiple replicates, simuPOP allows you to take full advantage of a multicore computer because it can be configured to run multithreaded (check the documentation). In this case, the more you depend on simuPOP native operators, the better. Python-coded operators will be single threaded because of Python's **Global Interpreter Lock** (**GIL**). If you want to know more about the GIL refer to `http://www.dabeaz.com/python/UnderstandingGIL.pdf`.

While performing multiple replicates of complex models, I prefer to use a different strategy, that is, I perform a single replicate per process, but run multiple processes. This has the advantage of scaling a cluster, whereas simuPOP multithreaded code can only use a single computer. For very complex simulations, I do not compute any statistics at all with simuPOP; I just save the results to a file (simuPOP has operators to dump data, for example, in Genepop format) and then use external applications to compute any statistics. Again, this strategy is just for very complex simulations with many replicates. If your requirements are simpler, multithreaded simuPOP with its built-in statistical methods will be enough.

Simulating population structure using island and stepping-stone models

We will now simulate population structure. Let's start with an island model and then create a one-dimensional stepping-stone model. We will also study F_{ST} and distinguish between deme-level statistics and meta-population level statistics. Strictly speaking, we will simulate fragmentation models by splitting into islands or stepping-stones.

Getting ready

Read the first recipe (*Introducing forward-time simulations*) as it introduces the basic programming framework. If you are using notebooks, the content is in `04_PopSim/Pop_Structure.ipynb`.

How to do it...

Take a look at the following steps:

1. Let's start with some basic code from the first recipe:

    ```
    from __future__ import division
    from collections import defaultdict, OrderedDict
    from copy import deepcopy
    import simuPOP as sp
    from simuPOP import demography
    ```

```
num_loci = 10
pop_size = 50
num_gens = 101
num_pops = 10
migs = [0, 0.005, 0.01, 0.02, 0.05, 0.1]
init_ops = OrderedDict()
pre_ops = OrderedDict()
post_ops = OrderedDict()
pops = sp.Population([pop_size] * num_pops, loci=[1] *
    num_loci, infoFields=['migrate_to'])
```

❑ Note that we will simulate an island model with 10 islands (num_pops). Also, we will introduce a new infoField, migrate_to, which is necessary to implement migration. We will try out several migration rates (including 0). Also, when we create a population, the population size is now a list of 10 values (we could have demes with different sizes, but we will keep them constant here).

2. We will include a variation of previous functions to accumulate values as follows:

```
def init_accumulators(pop, param):
    accumulators = param
    for accumulator in accumulators:
        if accumulator.endswith('_sp'):
            pop.vars()[accumulator] = defaultdict(list)
        else:
            pop.vars()[accumulator] = []
    return True
def update_accumulator(pop, param):
    accumulator, var = param
    if  var.endswith('_sp'):
        for sp in range(pop.numSubPop()):
            pop.vars()[accumulator][sp].append(
    deepcopy(pop.vars(sp)[var[:-3]]))
        else:
            pop.vars()[accumulator].append(deepcopy(pop.vars()[var]))
    return True
```

❑ simuPOP allows you to compute statistics per subpopulation. For example, if you have an island model with 10 populations, you can actually compute 11 allele frequencies per locus: 10 for each deme plus one for the meta population (that is, all 10 demes are considered as a single population). The preceding functions cater to this (as simuPOP variables for subpopulations are suffixed with _sp).

3. Let's add operators and run a simulation as follows:

```
init_ops['accumulators'] = sp.PyOperator(init_accumulators,
param=['fst'])
init_ops['Sex'] = sp.InitSex()
init_ops['Freq'] = sp.InitGenotype(freq=[0.5, 0.5])
for i, mig in enumerate(migs):
    post_ops['mig-%d' % i] = \
sp.Migrator(demography.migrIslandRates(mig, num_pops),
    reps=[i])
post_ops['Stat-fst'] = sp.Stat(structure=sp.ALL_AVAIL)
post_ops['fst_accumulation'] = \
sp.PyOperator(update_accumulator, param=('fst', 'F_st'))
mating_scheme = sp.RandomMating()
sim = sp.Simulator(pops, rep=len(migs))
sim.evolve(initOps=init_ops.values(),
preOps=pre_ops.values(), postOps=post_ops.values(),
            matingScheme=mating_scheme, gen=num_gens)
```

   □ We will compute F_{ST} over all loci here; there is nothing special about how this is done. We just add its statistical operator and support functions to accumulate its result over the generations. simuPOP supports many other statistic operators; be sure to check the manual.

   □ There is an operator to perform the island migration (`migrIslandRates`). This requires the number of demes and the migration rate (that is, the fraction of individuals that migrate).

   □ As you have seen in the previous recipe, simuPOP allows you to execute replicates with same parameters. However, you can also vary the parameters per replicate. This is what we do in this case, that is, we replicate 0 with a migration of 0, replicate 1 of 0.005, and so on. So, different replicates will simulate different things.

4. Let's plot F_{ST} over time for all different migration rates as follows:

```
import seaborn as sns
sns.set_style('white')
import matplotlib.pyplot as plt
fig = plt.figure(figsize=(16, 9))
ax = fig.add_subplot(111)
for i, pop in enumerate(sim.populations()):
    ax.plot(pop.dvars().fst, label='mig rate %.4f' % migs[i])
ax.legend(loc=2)
ax.set_ylabel('F_ST')
ax.set_xlabel('Generation')
```

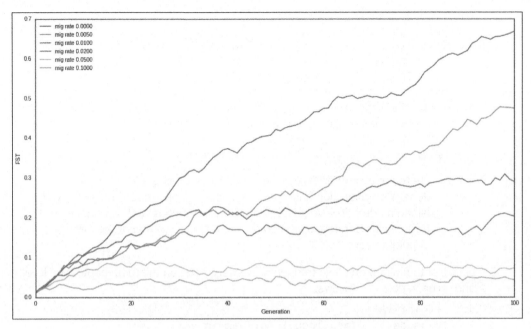

Figure 4: F_{ST} evolution over time with different migration rates in an island model with 10 demes, each with 50 individuals

□ We take the result from each replicate and plot it; each line will represent a different migration rate.

5. Let's now look at the stepping-stone model to study the behavior of population genetic parameters in the meta population and on each deme, let's start with some standard code:

```
num_gens = 400
num_loci = 5
init_ops = OrderedDict()
pre_ops = OrderedDict()
post_ops = OrderedDict()
init_ops['Sex'] = sp.InitSex()
init_ops['Freq'] = sp.InitGenotype(freq=[0.5, 0.5])
post_ops['Stat-freq'] = sp.Stat(alleleFreq=sp.ALL_AVAIL,
    vars=['alleleFreq', 'alleleFreq_sp'])
init_ops['accumulators'] = sp.PyOperator(init_accumulators,
    param=['allele_freq', 'allele_freq_sp'])
post_ops['freq_accumulation'] = \
sp.PyOperator(update_accumulator, param=('allele_freq',
    'alleleFreq'))
post_ops['freq_sp_accumulation'] = \
sp.PyOperator(update_accumulator, param=('allele_freq_sp',
    'alleleFreq_sp'))
```

```
for i, mig in enumerate(migs):
    post_ops['mig-%d' % i] =
sp.Migrator(demography.migrSteppingStoneRates(mig,
    num_pops), reps=[i])
pops = sp.Population([pop_size] * num_pops, loci=[1] *
    num_loci, infoFields=['migrate_to'])
sim = sp.Simulator(pops, rep=len(migs))
sim.evolve(initOps=init_ops.values(), preOps=pre_ops.values(),
    postOps=post_ops.values(),
    matingScheme=mating_scheme, gen=num_gens)
```

- ❑ There are two differences. One is that we are using a function to generate a migration based on the stepping-stone model instead of the island model. However, note that we are computing and storing allele frequencies per subpopulation (`alleleFreq_sp`) and at the meta population level (`alelleFreq`).

6. Let's plot the minimum allele frequency for every locus on the meta population level and on each deme as follows:

```
def get_maf(var):
    locus_data = [gen[locus] for gen in var]
    maf = [min(freq.values()) for freq in locus_data]
    maf = [v if v != 1 else 0 for v in maf]
    return maf


fig, axs = plt.subplots(3, num_pops // 2 + 1, figsize=(16,
    9), sharex=True, sharey=True, squeeze=False)
fig.suptitle('Minimum allele frequency at the meta-
population and 5 demes', fontsize='xx-large')
for line, pop in enumerate([sim.population(0),
sim.population(1), sim.population(len(migs) - 1)]):
    for locus in range(num_loci):
        maf = get_maf(pop.dvars().allele_freq)
        axs[line, 0].plot(maf)
        axs[line, 0].set_axis_bgcolor('black')
    for nsp in range(num_pops // 2):
        for locus in range(num_loci):
            maf = get_maf(pop.dvars().allele_freq_sp[nsp *
2])
            axs[line, nsp + 1].plot(maf)
fig.subplots_adjust(hspace=0, wspace=0)
```

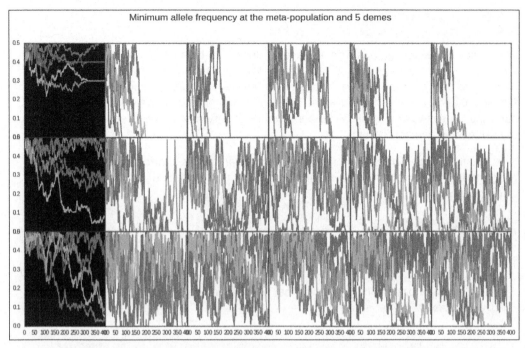

Figure 5: MAF in 5 loci with three different migration rates in the meta-population and 5 demes out of 10.

The preceding figure, *Figure 5*, is the MAF in 5 loci. The left-most chart (black background) represents the evolution at the meta-population level. The other 5 charts represent the evolution in 5 of the 10 demes in the stepping-stone model. The top line is without migration, the middle line with **0.005** migration and the bottom line with **0.1**. On the top line (no migration), note how different alleles can fixate on different demes, whereas at the meta level, the frequency is maintained at intermediate levels.

Modeling complex demographic scenarios

Here, we will show how simuPOP can be extremely flexible on demographic modeling. We will simulate an age-structured population with different fecundity per age for males and different maximum litter size for females. In the middle of the simulation, we will remove all older females. We will study the effective population size across the simulation. Furthermore, we will simulate multiallelic loci this time (for example, simulation of microsatellites).

The removal of a part of the population can model many things; for example, the management of a conserved population in a national park, or the usage of insecticides in vector populations, or modeling the illegal poaching of animals. The applications of this kind of modeling are plenty.

Getting ready

We will have three age groups. The first age group will not be able to reproduce (modeling infants). The two older age groups can. Males of age two have twice the chance of mating than males of age three. Females of age two can have many offspring in a cycle, whereas females of age three can only have one. Males of age one have 80 percent chance of surviving to age two and again 80 percent chance of surviving to age three. For females, the value is 90 percent for both.

Read the first recipe (*Introducing forward-time simulations*) as it introduces the basic programming framework. If you are using notebooks, this content is in `04_PopSim/Complex.ipynb`.

How to do it...

Take a look at the following steps:

1. Let's start by defining a function that will cull individuals according to its age and sex:

```
def kill(pop):
    kills = []
    for i in pop.individuals():
        if i.sex() == 1:
            cut = pop.dvars().survival_male[int(i.age)]
        else:
            cut = pop.dvars().survival_female[int(i.age)]
            if pop.dvars().gen > pop.dvars().cut_gen and \
i.age == 2:
                cut = 0
        if random.random() > cut:
            kills.append(i.ind_id)
    pop.removeIndividuals(IDs=kills)
    return True
```

   ❑ The function assumes that there are a couple of population variables (that we will create later) with the survival rate per sex. Also there is a provision to kill all females of age 2 after a certain generation.

2. We need to have a function to choose the parents as mating is far from random:

```
def choose_parents(pop):
    fathers = []
    mothers = []
    for ind in pop.individuals():
        if ind.sex() == 1:
```

```
            fathers.extend([ind] * pop.dvars().male_age_
    fecundity[int(ind.age)])
        else:
            ind.num_kids = 0
            mothers.append(ind)
    while True:
        father = random.choice(fathers)
        mother_ok = False
        while not mother_ok:
            mother = random.choice(mothers)
            if mother.num_kids < pop.dvars().max_kids[int(mother.
                age)]:
                mother.num_kids += 1
                mother_ok = True
        yield father, mother


def calc_demo(gen, pop):
    if gen > pop.dvars().cut_gen:
        add_females = len([ind for ind in
            pop.individuals([0, 2]) if ind.sex() == 2])
    else:
        add_females = 0
    return pop_size + pop.subPopSize([0, 3]) + add_females
```

- The `choose_parents` function will choose the father from a list that will include all males of age two and three. Males of age two get two entries (coming from a `male_age_fecundity` variable, which we will define later). Females will be allowed to have a maximum number of offspring according to their age.

- There is also a function to determine the next population size. This is mostly to maintain the population at a constant level. If we cull females of old age (as per the previous specification), we will use a virtual subpopulation (see step 4) that contains only old age individuals and get females.

3. The mating function is now a bit more complex than usual, as shown in the following code:

```
mating_scheme = sp.HeteroMating([
    sp.HomoMating(
        sp.PyParentsChooser(choose_parents),
        sp.OffspringGenerator(numOffspring=1, ops=[
            sp.MendelianGenoTransmitter(), sp.IdTagger()]),
        weight=1),
    sp.CloneMating(weight=-1)],
    subPopSize=calc_demo)
```

❏ Mating is now much more complex than the standard random mating; the `CloneMating` part will copy all individuals to the next cycle, whereas the `HomoMating` will add a few extra individuals according to the choice of parents in the preceding function.

4. Let's add some necessary boilerplate code:

```
pop_size = 300
num_loci = 50
num_alleles = 10
num_gens = 90
cut_gen = 50
max_kids = [0, 0, float('inf'), 1]
male_age_fecundity = [0, 0, 2, 1]
survival_male = [1, 0.8, 0.8, 0]
survival_female = [1, 0.9, 0.9, 0]
pops = sp.Population(pop_size, loci=[1] * num_loci,
    infoFields=['age', 'ind_id', 'num_kids'])
pops.setVirtualSplitter(sp.InfoSplitter(field='age',
    cutoff=[1, 2, 3]))
```

❏ Note the fecundity and survival variables and the new `infoFields` as well.

❏ However, the most important novelty is the creation of virtual subpopulations; simuPOP allows you to split your population into virtual subgroups according to many criteria. In our case, we will have virtual subpopulations divided by age. For example, we already used this to get older females of the population in the culling stage. Check simuPOP documentation on this concept because it's very powerful.

5. Let's create all operators and run the simulation as follows:

```
init_ops = OrderedDict()
pre_ops = OrderedDict()
post_ops = OrderedDict()
def init_age(pop):
    pop.dvars().male_age_fecundity = male_age_fecundity
    pop.dvars().survival_male = survival_male
    pop.dvars().survival_female = survival_female
    pop.dvars().max_kids = max_kids
    pop.dvars().cut_gen = cut_gen
    return True
def init_accumulators(pop, param):
    accumulators = param
    for accumulator in accumulators:
```

```
            pop.vars()[accumulator] = []
        return True
def update_pyramid(pop):
    pyr = defaultdict(int)
    for ind in pop.individuals():
        pyr[(int(ind.age), int(ind.sex()))] += 1
    pop.vars()['age_pyramid'].append(pyr)
    return True
def update_ldne(pop):
    pop.vars()['ldne'].append(pop.dvars().Ne_LD[0.05])
    return True
init_ops['Sex'] = sp.InitSex()
init_ops['ID'] = sp.IdTagger()
init_ops['accumulators'] = sp.PyOperator(init_accumulators,
    param=['ldne', 'age_pyramid'])
init_ops['Freq'] = sp.InitGenotype(freq=[1 / num_alleles] *
    num_alleles)
init_ops['Age-prepare'] = sp.PyOperator(init_age)
init_ops['Age'] = sp.InitInfo(lambda: random.randint(0,
    len(survival_male) - 1), infoFields='age')
pre_ops['Kill'] = sp.PyOperator(kill)
pre_ops['Age'] = sp.InfoExec('age += 1')
pre_ops['pyramid_accumulator'] = \
sp.PyOperator(update_pyramid)
post_ops['Ne'] = sp.Stat(effectiveSize=sp.ALL_AVAIL,
    subPops=[[0, 0]], vars=['Ne_LD'])
post_ops['Ne_accumulator'] = sp.PyOperator(update_ldne)
sim = sp.Simulator(pops, rep=1)
sim.evolve(initOps=init_ops.values(),
    preOps=pre_ops.values(), postOps=post_ops.values(),
    matingScheme=mating_scheme, gen=num_gens)
```

- We run the simulation for 90 generations and start killing females of age two at generation 50. We will consider the first 10 generations as burn-in.

- We will use markers with 10 alleles (microsatellite-like).

- Note that age pyramid functions used to store the evolution of age structure over time. More generally, note all code used to manipulate age.

- We compute an effective population size (Ne) estimator based on linkage disequilibrium. Note that we compute this only on virtual 0 subpopulation (that is, newborns). This is because the **Ne** estimation with this method only makes sense if you use a single cohort of individuals.

6. We can now extract all values and plot the Ne estimation over time along with the age pyramid:

```
ld_ne = sim.population(0).dvars().ldne
pyramid = sim.population(0).dvars().age_pyramid
sns.set_palette('Set2')

fig = plt.figure(figsize=(16, 9))
ax_ldne = fig.add_subplot(211)
ax_ldne.plot([x[0] for x in ld_ne[10:]])
ax_ldne.plot([x[1] for x in ld_ne[10:]], 'k--')
ax_ldne.plot([x[2] for x in ld_ne[10:]], 'k--')
ax_ldne.set_xticks(range(0, 81, 10))
ax_ldne.set_xticklabels([str(x) for x in range(10, 91, 10)])
ax_ldne.axvline(cut_gen - 10)
ax_ldne.set_xlabel('Cycle')
ax_ldne.set_ylabel('Effective population size (Est)')
def plot_pyramid(ax_bp, pyramids):
    bp_data = [([], []) for group in range(3)]
    for my_pyramid in pyramids:
        for (age, sex), cnt in my_pyramid.items():
            bp_data[age - 1][sex - 1].append(cnt)
    for group in range(3):
        bp = sns.boxplot(bp_data[group], positions=[group *
3 + 1, group * 3 + 2], widths=0.6, ax=ax_bp)
        ax_bp.text(1 + 3 * group, 90, 'M', va='top',
            ha='center')
        ax_bp.text(2 + 3 * group, 90, 'F', va='top',
            ha='center')
    ax_bp.set_xlim(0, 9)
    ax_bp.set_ylim(20, 90)
    ax_bp.set_xticklabels(['1', '2', '3'])
    ax_bp.set_xticks([1.5, 4.5, 7.5])
    ax_bp.legend()
pre_decline = pyramid[10:50]
post_decline = pyramid[51:]
ax_bp = fig.add_subplot(2, 2, 3)
plot_pyramid(ax_bp, pre_decline)
ax_bp = fig.add_subplot(2, 2, 4)
plot_pyramid(ax_bp, post_decline)
```

❏ The top chart shows the Ne estimation (including 5 and 95 percent confidence intervals in dashed lines).

❑   The bottom charts shows the distribution of number of individuals per age group (1 to 3) and sex. The left chart shows the version before the cull of old age females, whereas the right chart shows the version after the cull:

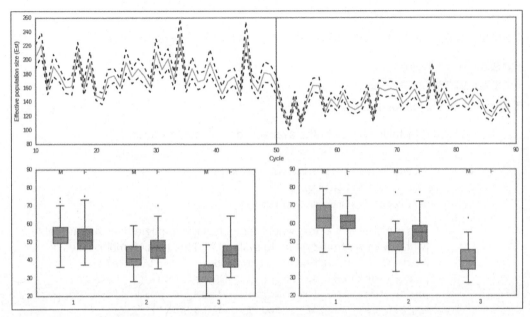

Figure 6: Top chart: Ne estimation over time and bottom charts: the age pyramid before culling (left) and after (right)

In preceding figure, *Figure 6*, the top chart depicts the change in effective population size estimation over time (the blue line depicts where old-aged females start to get culled). The bottom chart shows the distribution of number of individuals per age group (1, 2 or 3 years) and sex (males in pink and females in brown). The left chart is before the cull and the right chart is after the cull.

Simulating the coalescent with Biopython and fastsimcoal

While there is a native Python coalescent simulator called CoaSim, this is a very old application that does not run on modern Python versions. As such, if you want to do coalescent simulations with Python, Biopython offers a wrapper to fastsimcoal and the older simcoal.

Here, we will generate some configuration files for fastsimcoal with the demographic and genomic information. We will also run the coalescent simulator from Python.

Getting ready

You will need to obtain fastsimcoal from `http://cmpg.unibe.ch/software/fastsimcoal2/`. If you are using the Docker image, this is taken care of for you.

If you are using notebooks, this content is in `04_PopSim/Coalescent.ipynb`.

How to do it...

Take a look at the following steps:

1. Let's start by taking a look at the available demographic models:

```
import os
from Bio.PopGen import SimCoal as simcoal
print(simcoal.builtin_tpl_dir)
print(os.listdir(simcoal.builtin_tpl_dir))
```

- This will print a directory where templates for demographies are available (which will vary from system to system) and the list of template files inside. Currently, we have:`['simple.par', 'split_island. par', 'decline_split.par', 'bottle.par', 'ssm_1d. par', 'decline_lambda.par', 'ssm_2d.par', 'island.par', 'split_ssm_2d.par' 'split_ssm_1d.par'].`

- If you open any of the preceding files (I suggest starting with `simple.par`), you will see the template for fastsimcoal. All keywords starting with the `?` character need to be given as parameters. Thus, for any model that you want to use, you will need to check the template for parameters. Model names give an indication of the demography being modeled.

2. Let's create the necessary data structure to generate some demographies as follows:

```
simple = [('sample_size', [30]),
          ('pop_size', [100, 200])]
island = [('sample_size', [30]),
          ('pop_size', [100]),
          ('mig', [0.01]),
          ('total_demes', [10])]
split_ssm_1d = [('sample_size', [30]),
                ('pop_size', [100]),
                ('mig', [0.01]),
                ('ne', [500]),
                ('t', [100]),
                ('total_demes', [10])]
```

- ❏ The first demography is a "simple demography", that is, a single population with constant size. We will sample 30 individuals (remember that coalescent simulation normally deals with haploid data, but check the fastsimcoal manual for more details). Note that the population size is either 100 or 200. In this case, not one, but two template files will be generated, one for each population size. In theory, you can put more than one value for each parameter, but be aware that this may generate lots of combinations of files.

- ❏ Next comes the typical island model and finally, a one-dimensional stepping-stone model.

3. Now that we have the demographies, let's consider some genomes:

```
n_indep_MSATs = [(200, [('MICROSAT', [1, 0, 0.005, 0,
    0])])]
linked_snps = [(1, [('SNP', [200, 0.0005, 0.01])])]
linked_DNA = [(1, [('DNA', [1000, 0.0005, 0.0000002,
    0.33])])]
complex_genome = [(2, [
    ('DNA', [10, 0.00001, 0.00005, 0.33]),
    ('SNP', [1, 0.001, 0.0001]),
    ('MICROSAT', [1, 0.0, 0.001, 0.0, 0.0])])]
```

- ❏ The first case creates 200 chromosomes, each with one microsatellite with a mutation rate of 0.005. This is to say 200 unlinked microsatellites.

- ❏ The next line creates a single chromosome with 200 SNPs with a recombination rate of 0.0005 among them and a minimum allelele frequency of 10 percent.

- ❏ Then, we have a DNA sequence of 1000 nucleotides with a recombination rate of 0.0005 per base pair, a mutation rate of 0.0000002 per bp, and a transition rate of 0.33.

- ❏ Finally, we have a complex genome with two identical chromosomes, each with an assortment of markers. You can add more chromosomes to the list with whatever structure you see fit (that is, chromosomes can be different among themselves).

- ❏ For all possible types of markers and its parameters, check the fastsimcoal documentation.

4. We can now generate the (fast)simcoal parameter file:

```
from Bio.PopGen.SimCoal.Template import
generate_simcoal_from_template
generate_simcoal_from_template('island', complex_genome,
    island)
```

```
generate_simcoal_from_template('simple', n_indep_MSATs,
    simple)
generate_simcoal_from_template('split_ssm_1d', linked_snps,
    split_ssm_1d)
```

- ❑ This template generator requires the model name from step 1 (removing the `.par` prefix), followed by the genomic and demographic structure.

- ❑ This will generate the parameter file with the name based on the template and the parameter, for example, the template for the simple model will generate two model files. Remember that we have a population size of 100 and 200: `simple_100_30.par` and `simple_100_30.par`.

- ❑ This means that if you have the same demographic model with different genomic parameters, you will have to rename one of the parameter files before creating the second one because this will be overwritten.

5. Finally, we can run fastsimcoal, as shown in the following code:

```
from Bio.PopGen.SimCoal.Controller import \
FastSimCoalController as fsc
ctrl = fsc(bin_name='fsc251')
ctrl.run_fastsimcoal('simple_100_30.par', 10)
ctrl.run_fastsimcoal('simple_200_30.par', 10)
ctrl.run_fastsimcoal('split_ssm_1d_10_100_500_0.01_100_30.
par', 10)
ctrl.run_fastsimcoal('island_10_0.01_100_30.par', 10)
```

- ❑ The binary name is optional and is the current default on Biopython. In the probable case that there is a new fastsimcoal version with a new name, do not forget to change this.

- ❑ We will run 10 replicates for each parameter file. The result computed by fastsimcoal will be available in a directory with the name of the parameter file (minus the `.par` suffix). Take a look at the output created in those directories; they are in the Arlequin format.

There's more...

While there are Python facilities to generate input files for `(fast)simcoal` and run the application, there are currently no Python-based parsers for the Arlequin format generated by both these coalescent simulators. This means that there is a non-Python step to convert the Arlequin data to another format. Note that the next format is not clear-cut. If you are using SNPs or microsatellites, then Genepop may be an option, but if you are using DNA sequences, you will have to consider alternatives. If you have a single population, FASTA will probably suffice, but if you have a complex genome simulation, you may have to check which format best suits you. In a worst-case scenario, you may want to use the Arlequin application to perform the data analysis that is completely outside Python. Arlequin 3.5 has a command-line version that may be possible to automate.

In theory, fastsimcoal should be a faster version of simcoal, but you may want to compare the results of one against the other, especially if you are using very complex genomes. Biopython still has a controller for the old version.

See also

If you are interested in the coalescent simulation and data conversion, check the following links:

- The program fastsimcoal can be found at `http://cmpg.unibe.ch/software/fastsimcoal2/`.

- The old simcoal is still available at `http://cmpg.unibe.ch/software/simcoal2/`.

- Arlequin is widely used population genetics software capable of reading the output from fastsimcoal, it can be found at `http://cmpg.unibe.ch/software/arlequin35/`.

- For more documentation on the Biopython support for the coalescent simulation, check the Biopython tutorial at `http://biopython.org/DIST/docs/tutorial/Tutorial.html`.

- To convert Arlequin data, there are several options. GeneAlEx (`http://biology-assets.anu.edu.au/GenAlEx/Welcome.html`) does this and much more. If you are on Windows or Mac and have Excel available, you can consider it as a tool for population genetics analysis.

6
Phylogenetics

In this chapter, we will cover the following recipes:

- ▸ Preparing the Ebola dataset
- ▸ Aligning genetic and genomic data
- ▸ Comparing sequences
- ▸ Reconstructing phylogenetic trees
- ▸ Playing recursively with trees
- ▸ Visualizing phylogenetic data

Introduction

Phylogenetics is the application of molecular sequencing to study the evolutionary relationship among organisms. The typical way to illustrate this process is through the use of phylogenetic trees. The computation of these trees from genomic data is an active field of research with many real-world applications.

We will take the practical approach mentioned in this book to a new level; most of the recipes here are inspired by a very recent study on the Ebola virus, which researched the recent Ebola outbreak in Africa. This study is called *Genomic surveillance elucidates Ebola virus origin and transmission during the 2014 outbreak* from Gire et al published on Science and is available at http://www.sciencemag.org/content/345/6202/1369.short. Here, we will try to follow a similar methodology to arrive at similar results from the paper.

In this chapter, we will use **DendroPy** (a phylogenetics library) and Biopython. We will use DendroPy Version 3. Version 4 may be out when you read this. Most of this code will only work on Python 2.

Preparing the Ebola dataset

Here, we will download and prepare the dataset to be used for our analysis. The dataset contains complete genomes of the Ebola virus. We will use DendroPy to download and prepare the data.

Getting ready

We will download complete genomes from Genbank; these genomes were collected from various Ebola outbreaks, including several from the 2014 outbreak. Note that there are several virus species that cause the Ebola virus disease; the species involved in the 2014 outbreak (the EBOV virus, formally known as the Zaire Ebola virus) is the most common, but this disease is caused by more species of the genus *Ebola virus*; four others are also available in sequenced form. You can read more at `https://en.wikipedia.org/wiki/Ebolavirus`. For scientific references of all downloaded entries, check the GenBank records.

If you have dealt with previous chapters, you may panic looking at the potential data sizes involved here; this is not a problem at all because these are genomes of virus of around 19 kbp each. So, our approximately 100 genomes are actually quite light.

As usual, this information is available in the corresponding notebook available at `05_Phylo/Exploration.ipynb`.

How to do it...

Take a look at the following steps:

1. First, let's start specifying our data sources using DendroPy as follows:

```
from __future__ import division, print_function
import dendropy
from dendropy.interop import genbank
def get_ebov_2014_sources():
    #EBOV_2014
    #yield 'EBOV_2014', \
genbank.GenBankDna(id_range=(233036, 233118), prefix='KM')
    yield 'EBOV_2014', genbank.GenBankDna(id_range=(34549,
        34563), prefix='KM0')


def get_other_ebov_sources():
    #EBOV other
    yield 'EBOV_1976', genbank.GenBankDna(ids=['AF272001',
        'KC242801'])
    yield 'EBOV_1995', genbank.GenBankDna(ids=['KC242796',
        'KC242799'])
```

```
        yield 'EBOV_2007', genbank.GenBankDna(id_range=(84,
            90), prefix='KC2427')


def get_other_ebolavirus_sources():
    #BDBV
    yield 'BDBV', genbank.GenBankDna(id_range=(3, 6),
        prefix='KC54539')
    yield 'BDBV', genbank.GenBankDna(ids=['FJ217161'])


    #RESTV
    yield 'RESTV', genbank.GenBankDna(ids=['AB050936',
        'JX477165', 'JX477166', 'FJ621583', 'FJ621584',
        'FJ621585'])


    #SUDV
    yield 'SUDV', genbank.GenBankDna(ids=['KC242783',
        'AY729654', 'EU338380',
                                        'JN638998',
        'FJ968794', 'KC589025', 'JN638998'])
    #yield 'SUDV', genbank.GenBankDna(id_range=(89, 92),
prefix='KC5453')


    #TAFV
    yield 'TAFV', genbank.GenBankDna(ids=['FJ217162'])
```

- ❏ We have three functions; one to retrieve data from the most recent EBOV outbreak, another from previous EBOV outbreaks, and one from outbreaks of other species.

- ❏ Note that the DendroPy GenBank interface provides several different ways to specify lists or ranges of records to retrieve.

- ❏ Some lines are commented out. These include code to download more genomes. For our purpose, the subset that we will download is enough.

2. Now, we will create a set of FASTA files; we will use these files here and in future recipes with subsets of data to analyze:

```
other = open('other.fasta', 'w')
sampled = open('sample.fasta', 'w')

for species, recs in get_other_ebolavirus_sources():
    char_mat = \
recs.generate_char_matrix(taxon_set=dendropy.TaxonSet(),
        gb_to_taxon_func=lambda gb: dendropy.Taxon(label='%s_%s' %
            (species, gb.accession)))
    char_mat.write_to_stream(other, 'fasta')
    char_mat.write_to_stream(sampled, 'fasta')
```

```
other.close()
ebov_2014 = open('ebov_2014.fasta', 'w')
ebov = open('ebov.fasta', 'w')
for species, recs in get_ebov_2014_sources():
    char_mat = recs.generate_char_matrix(taxon_set=dendropy.
        TaxonSet(),
        gb_to_taxon_func=lambda gb: dendropy.
Taxon(label='EBOV_2014_%s' % gb.accession))
    char_mat.write_to_stream(ebov_2014, 'fasta')
    char_mat.write_to_stream(sampled, 'fasta')
    char_mat.write_to_stream(ebov, 'fasta')
ebov_2014.close()

ebov_2007 = open('ebov_2007.fasta', 'w')
for species, recs in get_other_ebov_sources():
    char_mat = recs.generate_char_matrix(taxon_set=dendropy.
TaxonSet(),
        gb_to_taxon_func=lambda gb: dendropy.Taxon(label='%s_%s' %
                (species, gb.accession)))
    char_mat.write_to_stream(ebov, 'fasta')
    char_mat.write_to_stream(sampled, 'fasta')
    if species == 'EBOV_2007':
        char_mat.write_to_stream(ebov_2007, 'fasta')

ebov.close()
ebov_2007.close()
sampled.close()
```

- ❏ We will generate several different FASTA files, which include all genomes, just EBOV, or just EBOV samples from the 2014 outbreak. In this chapter, we will mostly use the `sample.fasta` file with all genomes. You can find a detailed description of DendroPy's data structures in its documentation.

- ❏ Note that the use of DendroPy functions to create FASTA files retrieved GenBank records are converted. Also note that the ID of each sequence on the FASTA file is produced by a lambda function that uses species and year apart from the GenBank accession.

3. Let's extract four (of the total of seven) genes in the virus as follows:

```
my_genes = ['NP', 'L', 'VP35', 'VP40']

def dump_genes(species, recs, g_dls, p_hdls):
    for rec in recs:
        for feature in rec.feature_table:
                if feature.key == 'CDS':
```

```
                        gene_name = None
                        for qual in feature.qualifiers:
                            if qual.name == 'gene':
                                if qual.value in my_genes:
                                    gene_name = qual.value
                            elif qual.name == \
                                    'translation':
                                protein_translation = \
                                    qual.value
                        if gene_name is not None:
                            locs = \
                                feature.location.split('.')
                            start, end = int(locs[0]),
                                int(locs[-1])
                            g_hdls[gene_name].write('>%s_%s\n' %
                                (species, rec.accession))
                            p_hdls[gene_name].write('>%s_%s\n' %
                                (species, rec.accession))
                            g_hdls[gene_name].write('%s\n' %
                                rec.sequence_text[start - 1 :
                                    end])
                            p_hdls[gene_name].write('%s\n' %
                                protein_translation)


g_hdls = {}
p_hdls = {}
for gene in my_genes:
    g_hdls[gene] = open('%s.fasta' % gene, 'w')
    p_hdls[gene] = open('%s_P.fasta' % gene, 'w')
for species, recs in get_other_ebolavirus_sources():
    if species in ['RESTV', 'SUDV']:
        dump_genes(species, recs, g_hdls, p_hdls)
for gene in my_genes:
    g_hdls[gene].close()
    p_hdls[gene].close()
```

❏ We start by searching the first GenBank record for all gene features (refer to *Chapter 2, Next-generation Sequencing,* or the NCBI documentation for further details; although we will use DendroPy and not Biopython here, the concepts are similar) and write to FASTA files in order to extract the genes. We put each gene in a different file and only take two virus species.

❏ We also get translated proteins; which are available on the records for each gene.

4. Let's create a function to get the basic statistical information from the alignment as follows:

```
def describe_seqs(seqs):
    print('Number of sequences: %d' % len(seqs.taxon_set))
    print('First 10 taxon sets: %s' % ' '.join([taxon.label
for taxon in seqs.taxon_set[:10]]))
    lens = []
    for tax, seq in seqs.items():
        lens.append(len([x for x in seq.symbols_as_list()
            if x != '-']))
    print('Genome length: min %d, mean %.1f, max %d' %
        (min(lens), sum(lens) / len(lens), max(lens)))
```

- ❑ Our function takes a `DendroPy` class (`DnaCharacterMatrix`) and counts the number of taxons. We then extract all amino acids per sequence (we exclude gaps identified by `-`) to compute the length, and report the minimum, mean, and maximum sizes. Take a look at the DendroPy documentation for details on the API.

5. Let's inspect the sequence of the EBOV genome and compute basic statistics as shown earlier:

```
ebov_seqs = \
    dendropy.DnaCharacterMatrix.get_from_path('ebov.fasta',
        schema='fasta', data_type='dna')
print('EBOV')
describe_seqs(ebov_seqs)
del ebov_seqs
```

- ❑ We then call a function and get 25 sequences with a minimum size of 18,613, mean of 18,909.8, and maximum of 18,959. A small genome when compared with eukaryotes.

- ❑ Note that at the very end, the memory structure is deleted. This is because the memory footprint is still quite big (DendroPy is a pure Python library and has some costs in terms of speed and memory). Be careful with your memory usage when you load full genomes.

6. Now, let's inspect the other *Ebola virus* genome file and count the number of different species:

```
print('ebolavirus sequences')
ebolav_seqs = \
    dendropy.DnaCharacterMatrix.get_from_path('other.fasta',
        schema='fasta', data_type='dna')
describe_seqs(ebolav_seqs)
from collections import defaultdict
species = defaultdict(int)
```

```
for taxon in ebolav_seqs.taxon_set:
    toks = taxon.label.split('_')
    my_species = toks[0]
    if my_species == 'EBOV':
        ident = '%s (%s)' % (my_species, toks[1])
    else:
        ident = my_species
    species[ident] += 1
for my_species, cnt in species.items():
    print("%20s: %d" % (my_species, cnt))
del ebolav_seqs
```

❏ The name prefix of each taxon is indicative of the species and we leverage that to fill a dictionary of counts.

❏ The output for species and the EBOV breakdown is as follows (with the legend as `Bundibugyo virus=BDBV`, `Tai Forest virus=TAFV`, `Sudan virus=SUDV`, and `Reston virus=RESTV`. We have 1 TAFV, 6 SUDV, 6 RESTV, and 5 BDBV.

7. Let's extract the basic statistics of a gene on the virus:

```
gene_length = {}
my_genes = ['NP', 'L', 'VP35', 'VP40']

for name in my_genes:
    gene_name = name.split('.')[0]
    seqs = dendropy.DnaCharacterMatrix.get_from_path('%s.fasta' %
        name, schema='fasta', data_type='dna')
    gene_length[gene_name] = []
    for tax, seq in seqs.items():
        gene_length[gene_name].append(len([x for x in
            seq.symbols_as_list() if x != '-']))
for gene, lens in gene_length.items():
    print ('%6s: %d' % (gene, sum(lens) / len(lens)))
```

This allows you to have an overview of the basic gene information (name and mean size) as follows:

```
NP: 2218
VP40: 988
L: 6636
VP35: 990
```

There's more...

Most of the work here can probably be performed with Biopython, but DendroPy has additional functionalities that will be explored in later recipes. Furthermore, as you will see, it's more robust with certain tasks (such as file parsing).

Most importantly, there is another Python library to perform phylogenetics that you should consider. It's called ETE and is available at `http://etetoolkit.org/`.

See also

▸ Wikipedia has a good introductory page on the Ebola virus disease at `http://en.wikipedia.org/wiki/Ebola_virus_disease`

▸ The wiki page about the virus is `http://en.wikipedia.org/wiki/Ebola_virus`; also see the page on the genus at `http://en.wikipedia.org/wiki/Ebolavirus`

▸ The reference application in phylogenetics is Joe Felsenstein's Phylip `http://evolution.genetics.washington.edu/phylip.html`.

▸ We will use the Nexus and Newick formats in future recipes (`http://evolution.genetics.washington.edu/phylip/newicktree.html`), but also check the PhyloXML format (`http://en.wikipedia.org/wiki/PhyloXML`)

Aligning genetic and genomic data

Before we can perform any phylogenetic analysis, we need to align our genetic and genomic data. Here, we will use MAFFT (`http://mafft.cbrc.jp/alignment/software/`) to perform the genome analysis and the gene analysis will be performed using MUSCLE (`http://www.drive5.com/muscle/`).

Getting ready

To perform the genomic alignment, you will need to install MAFFT, and to perform the genic alignment, MUSCLE will be used. Also, we will use TrimAl (`http://trimal.cgenomics.org/`) to remove spurious sequences and poorly aligned regions in an automated manner. On Ubuntu and Linux, MAFFT and MUSCLE can be installed using `apt-get install mafft muscle` packages. TrimAl will have to be manually installed.

As usual, this information is available in the corresponding notebook at `05_Phylo/Alignment.ipynb`. You will need to have run the previous notebook as it will generate files that are required here.

In this chapter, we will use Biopython.

How to do it...

Take a look at the following steps:

1. We will now run MAFFT to align genomes, as shown in the following code. This task is CPU-intensive and memory-intensive and will take quite some time:

```
from Bio.Align.Applications import MafftCommandline
mafft_cline = MafftCommandline(input='sample.fasta',
    ep=0.123, reorder=True, maxiterate=1000, localpair=True)
print(mafft_cline)
stdout, stderr = mafft_cline()
with open('align.fasta', 'w') as w:
    w.write(stdout)
```

   ❑ The parameters are the same as the one specified in the supplementary material of the paper. We will use the BioPython interface to call MAFFT.

2. Let's use TrimAl to trim sequences as follows:

```
os.system('./trimal -automated1 -in align.fasta -out
trim.fasta -fasta')
```

   ❑ Here we just call the application using `os.system`. The `-automated1` parameter is from the supplementary material.

3. We can also run MUSCLE to align proteins:

```
from Bio.Align.Applications import MuscleCommandline

my_genes = ['NP', 'L', 'VP35', 'VP40']

for gene in my_genes:
    muscle_cline = MuscleCommandline(input='%s_P.fasta' %
        gene)
    print(muscle_cline)
    stdout, stderr = muscle_cline()
    with open('%s_P_align.fasta' % gene, 'w') as w:
        w.write(stdout)
```

   ❑ Again, we will use Biopython to call an external application. Here, we will align a set of proteins.

   ❑ Note that to make some analysis of molecular evolution, we have to compare aligned genes, not proteins (for example, compare synonymous to nonsynonymous mutations). However, we have just aligned proteins. So, we have to "convert" the alignment to the gene sequence form.

4. Let's align the genes by finding three nucleotides that correspond to each amino acid:

```
from Bio import SeqIO
from Bio.Seq import Seq
from Bio.SeqRecord import SeqRecord
from Bio.Alphabet import generic_protein

for gene in my_genes:
    gene_seqs = {}
    unal_gene = SeqIO.parse('%s.fasta' % gene, 'fasta')
    for rec in unal_gene:
        gene_seqs[rec.id] = rec.seq

    al_prot = SeqIO.parse('%s_P_align.fasta' % gene,
        'fasta')
    al_genes = []
    for protein in al_prot:
        my_id = protein.id
        seq = ''
        pos = 0
        for c in protein.seq:
            if c == '-':
                seq += '---'
            else:
                seq += str(gene_seqs[my_id][pos:pos + 3])
                pos += 3
        al_genes.append(SeqRecord(Seq(seq), id=my_id))
    SeqIO.write(al_genes, '%s_align.fasta' % gene, 'fasta')
```

❑ The code gets the protein and the gene coding; if a gap is found in a protein, three gaps are written; if an amino acid is found, corresponding nucleotides of the gene are written.

Comparing sequences

Here, we will compare aligned sequences. We will perform gene and genome-wide comparisons.

Getting ready

We will use DendroPy and will require results from the previous two recipes. As usual, this information is available in the corresponding notebook at `05_Phylo/Comparison.ipynb`.

How to do it...

Take a look at the following steps:

1. Let's start analyzing the gene data. For simplicity, we will only use the data from two other species of the genus *Ebola virus* that are available in the extended dataset: the *Reston* virus (RESTV) and the *Sudan* virus (SUDV):

```
from __future__ import print_function
import os
from collections import OrderedDict
import dendropy
from dendropy import popgenstat
genes_species = OrderedDict()
my_species = ['RESTV', 'SUDV']
my_genes = ['NP', 'L', 'VP35', 'VP40']


for name in my_genes:
    gene_name = name.split('.')[0]
    char_mat = \
dendropy.DnaCharacterMatrix.get_from_path('%s_align.fasta'
    % name, 'fasta')
    genes_species[gene_name] = {}


    for species in my_species:
        genes_species[gene_name][species] = \
            dendropy.DnaCharacterMatrix()
    for taxon, char_map in char_mat.items():
        species = taxon.label.split('_')[0]
        if species in my_species:
            genes_species[gene_name][species].extend_map({taxon:
                char_map})
```

   ❑ We get four genes that we stored in the first recipe and aligned in the second.

   ❑ We load all the files (which are FASTA formatted) and create a dictionary with all the genes. Each entry will be a dictionary itself with the RESTV or SUDV species, including all reads. This is not a lot of data, just a handful of genes.

2. Let's print some basic information for all four genes, such as number of segregating sites, nucleotide diversity, Tajima's D, and Waterson's Theta (check the *See also* section of this recipe for links on these statistics):

```
import numpy as np
import pandas as pd
summary = np.ndarray(shape=(len(genes_species), 4 *
    len(my_species)))
```

```
stats = ['seg_sites', 'nuc_div', 'taj_d', 'wat_theta']
for row, (gene, species_data) in
    enumerate(genes_species.items()):
    for col_base, species in enumerate(my_species):
        summary[row, col_base * 4] = \
popgenstat.num_segregating_sites(species_data[species])
        summary[row, col_base * 4 + 1] = \
popgenstat.nucleotide_diversity(species_data[species])
        summary[row, col_base * 4 + 2] = \
popgenstat.tajimas_d(species_data[species])
        summary[row, col_base * 4 + 3] = \
popgenstat.wattersons_theta(species_data[species])
columns = []
for species in my_species:
    columns.extend(['%s (%s)' % (stat, species) for stat in
        stats])
df = pd.DataFrame(summary, index=genes_species.keys(),
    columns=columns)
df # vs print(df)
```

3. Let's look at the output first and then explain how to build it:

	seg_sites (RESTV)	nuc_div (RESTV)	taj_d (RESTV)	wat_theta (RESTV)	seg_sites (SUDV)	nuc_div (SUDV)	taj_d (SUDV)	wat_theta (SUDV)
NP	113	0.020659	-0.482275	49.489051	118	0.029630	1.203522	56.64
L	288	0.018143	-0.295386	126.131387	282	0.024193	1.412350	135.36
VP35	42	0.017099	-0.530330	18.394161	50	0.027761	1.069061	24.00
VP40	61	0.026155	-0.188135	26.715328	41	0.023517	1.269160	19.68

- I used a pandas DataFrame to print the results because it's really tailored to deal with an operation like this.

- We will initialize our DataFrame with a NumPy multidimensional array with four rows (genes) and four statistics times the two species.

- Statistics, such as number of segregating sites, nucleotide diversity, Tajima's D, and Watterson's Theta, are computed by DendroPy. Note the placement of individual data points in an array (the coordinate computation).

- Look at the very last line: if you are on the IPython Notebook, just putting the `df` at the end will render the DataFrame and cell output as well. If you are not on a notebook, perform a `print(df)` (you can also perform this in a notebook, but it will not look as pretty).

4. Let's now extract similar information, but genome-wide instead of only gene-wide. In this case, we will use a subsample of two EBOV outbreaks (from 2007 and 2014). We will perform a function to display basic statistics as follows:

```python
def do_basic_popgen(seqs):
    num_seg_sites = popgenstat.num_segregating_sites(seqs)
    avg_pair = popgenstat.average_number_of_pairwise_
differences(seqs)
    nuc_div = popgenstat.nucleotide_diversity(seqs)
    print('Segregating sites: %d, Avg pairwise diffs: %.2f,
Nucleotide diversity %.6f' % (num_seg_sites, avg_pair,
    nuc_div))
    print("Watterson's theta: %s" % popgenstat.wattersons_
theta(seqs))
    print("Tajima's D: %s" % popgenstat.tajimas_d(seqs))
```

❑ By now, this function should be easy to understand, given the preceding examples.

5. Let's now extract a subsample of the data properly and output the statistical information:

```python
ebov_seqs = \
dendropy.DnaCharacterMatrix.get_from_path('trim.fasta',
    schema='fasta', data_type='dna')
sl_2014 = []
drc_2007 = []
ebov2007_set = dendropy.DnaCharacterMatrix()
ebov2014_set = dendropy.DnaCharacterMatrix()
for taxon, char_map in ebov_seqs.items():
        if taxon.label.startswith('EBOV_2014') and \
        len(sl_2014) < 8:
        sl_2014.append(char_map)
        ebov2014_set.extend_map({taxon: char_map})
    elif taxon.label.startswith('EBOV_2007'):
        drc_2007.append(char_map)
        ebov2007_set.extend_map({taxon: char_map})
del ebov_seqs
print('2007 outbreak:')
print('Number of individuals: %s' %
    len(ebov2007_set.taxon_set))
do_basic_popgen(ebov2007_set)
print('\n2014 outbreak:')
print('Number of individuals: %s' %
    len(ebov2014_set.taxon_set))
do_basic_popgen(ebov2014_set)
```

- ❑ Here, we will construct two versions of two datasets: the 2014 outbreak and the 2007 outbreak. We generate a version as a `DnaCharacterMatrix` and another as a list. We will use this list version at the end of this recipe.

- ❑ As the dataset for the EBOV outbreak of 2014 is large, we subsample it with just eight individuals, a comparable sample size as the dataset of the 2007 outbreak.

- ❑ Again, we delete the `ebov_seqs` data structure to conserve memory (these are genomes, not only genes).

- ❑ If you perform this analysis on the complete dataset for the 2014 outbreak available on GenBank (99 samples), be prepared to wait for quite some time.

- ❑ The output is shown here:

```
2007 outbreak:
Number of individuals: 7
Segregating sites: 25, Avg pairwise diffs: 7.71, Nucleotide diversity 0.000412
Watterson's theta: 10.2040816327
Tajima's D: -1.38311415748

2014 outbreak:
Number of individuals: 8
Segregating sites: 6, Avg pairwise diffs: 2.79, Nucleotide diversity 0.000149
Watterson's theta: 2.31404958678
Tajima's D: 0.950120802758
```

6. Finally, we perform some statistical analysis on two subsets of 2007 and 2014 as follows:

```
pair_stats = \
popgenstat.PopulationPairSummaryStatistics(sl_2014,
    drc_2007)
print('Average number of pairwise differences irrespective
of population: %.2f' %
    pair_stats.average_number_of_pairwise_differences)
print('Average number of pairwise differences between
populations: %.2f' %
    pair_stats.average_number_of_pairwise_differences_between)
print('Average number of pairwise differences within
populations: %.2f' %
    pair_stats.average_number_of_pairwise_differences_within)
```

```
print('Average number of net pairwise differences : %.2f' %
        pair_stats.average_number_of_pairwise_differences_net)
print('Number of segregating sites: %d' %
        pair_stats.num_segregating_sites)
print("Watterson's theta: %.2f" %
        pair_stats.wattersons_theta)
print("Wakeley's Psi: %.3f" % pair_stats.wakeleys_psi)
print("Tajima's D: %.2f" % pair_stats.tajimas_d)
```

❏ Note that we will perform something slightly different here; we will ask DendroPy (`popgenstat.PopulationPairSummaryStatistics`) to directly compare two populations so that we get the following results:

```
Average number of pairwise differences irrespective of population: 284.46
Average number of pairwise differences between populations: 529.07
Average number of pairwise differences within populations: 10.50
Average number of net pairwise differences : 518.57
Number of segregating sites: 549
Watterson's theta: 168.84
Wakeley's Psi: 0.296
Tajima's D: 3.05
```

❏ The number of segregating sites is now much bigger because we are dealing with data from two different populations that are reasonably diverged.

❏ The average number of pairwise differences among populations is quite large. As expected, this is much larger than the average number of population irrespective of the population information.

There's more...

If you want to get many population genetics formulas, including the ones used here; I strongly recommend that you get the manual of the Arlequin software suite (`http://cmpg.unibe.ch/software/arlequin35/`). If you do not use Arlequin to perform data analysis, its manual is probably the best reference to implement formulas. This free document has probably more relevant formula implementation details than any book that I remember.

Reconstructing phylogenetic trees

Here, we will construct phylogenetic trees for the aligned dataset for all *Ebola* species. We will follow the procedure quite similar to the one used in the paper.

Getting ready

This recipe requires RAxML, a program for maximum likelihood-based inference of large phylogenetic trees, which you can check at `http://sco.h-its.org/exelixis/software.html`. With Ubuntu Linux, you can simply `apt-get install raxml`. Note that the binary is called `raxmlHPC`.

The code here is simple, but it will take time to execute because it will call RAxML (which is computationally intensive). If you opt to use the DendroPy interface, it may also become memory-intensive. We will interact with RAxML via DendroPy and Biopython, leaving you with a choice of which interface to use; DendroPy gives an easy way to access results, whereas Biopython is less memory-intensive. Although there is a recipe for visualization later in this chapter, we will nonetheless plot one of our generated trees here.

As usual, this information is available in the corresponding notebook at `05_Phylo/Reconstruction.ipynb`. You will need the output of the previous recipe.

How to do it...

Take a look at the following steps:

1. For DendroPy, we will load the data first and then reconstruct the genus dataset as follows:

```
from __future__ import print_function
import os
import shutil
import dendropy
from dendropy.interop import raxml
ebola_data = \
dendropy.DnaCharacterMatrix.get_from_path('trim.fasta',
    'fasta')
rx = raxml.RaxmlRunner()
ebola_tree = rx.estimate_tree(ebola_data, ['-m',
    'GTRGAMMA', '-N', '10'])
print('RAxML temporary directory %s:' %
    rx.working_dir_path)
del ebola_data
```

- ❏ Remember that the size of the data structure for this is quite big; therefore, be sure to have enough memory to load this.

- ❏ Be prepared to wait some time. Depending on your computer, this can be more than one hour. If it takes much longer, consider restarting the process as a RAxML bug might sometimes occur.

- ❏ We will run RAxML with the GTRΓ nucleotide substitution model as specified in the paper. We will only perform 10 replicates to speed up results, but you should probably do a lot more, say 100.

- ❏ At the end, we delete the genome data from memory as it takes up a lot of memory.

- ❏ The `ebola_data` variable will have the best RAxML tree with distances included.

- ❏ The `RaxmlRunner` object will have access to other information generated by RAxML.

- ❏ Let's print a directory where DendroPy will execute RAxML. If you inspect this directory, you will find a lot of files. As RAxML returns the best tree, you may want to ignore all these files, but we will discuss this a little in the Biopython alternative step.

2. We will save trees for future analysis; in our case, it will be visualization, as shown in the following code:

```
ebola_tree.write_to_path('my_ebola.nex', 'nexus')
```

- ❏ We will write sequences to a Nexus file because we need to store the topology information. FASTA is not enough here.

3. Let's visualize our genus tree as follows:

```
import matplotlib.pyplot as plt
from Bio import Phylo
my_ebola_tree = Phylo.read('my_ebola.nex', 'nexus')
my_ebola_tree.name = 'Our Ebolavirus tree'
fig = plt.figure(figsize=(16, 18))
ax = fig.add_subplot(1, 1, 1)
Phylo.draw(my_ebola_tree, axes=ax)
```

- ❏ We will defer the explanation of this code until the proper recipe further down the road, but if you look at the figure and compare it with the results from the paper, you will easily see that it looks a step in the right direction, for example, all individuals from the same species are clustered together.

□ You will notice that TrimAl changed names of its sequences. For example, adding their sizes. This is easy to solve; we will deal with this in the visualization recipe:

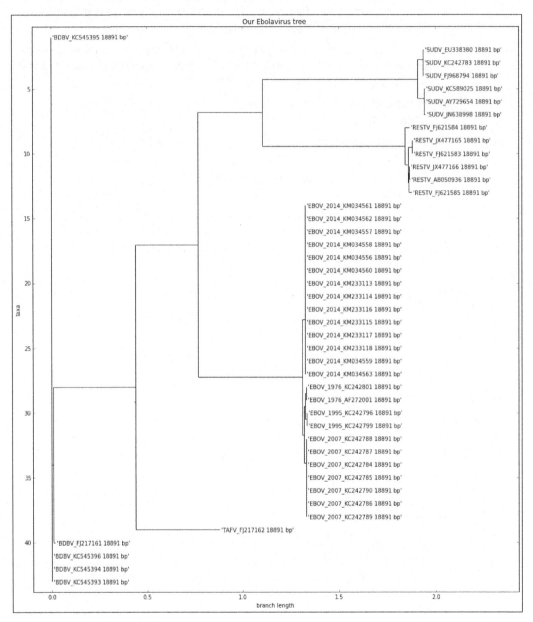

Figure 1: The phylogenetic tree that we generated with RAxML for all Ebola viruses

4. Let's reconstruct the phylogenetic tree with RAxML via Biopython. The Biopython interface is less declarative, but much more memory-efficient than DendroPy, so after running it, it will be your responsibility to process the output, whereas DendroPy automatically returns the best tree, as shown in the following code:

```
import random
import sys
from Bio.Phylo.Applications import RaxmlCommandline
raxml_cline = RaxmlCommandline(sequences='trim.fasta',
                               model='GTRGAMMA',
                               name='biopython',
                               num_replicates='10',
                               parsimony_seed=random.randint(0,
                               sys.maxint),
                               working_dir=os.getcwd() +
                               os.sep + 'bp_rx')
print(raxml_cline)
try:
        os.mkdir('bp_rx')
except OSError:
        shutil.rmtree('bp_rx')
        os.mkdir('bp_rx')
out, err = raxml_cline()
```

   ❑ DendroPy has a more declarative interface than Biopython, so you can take care of a few extra things. You should specify the seed (Biopython will put a fixed default of 10000 if you do not do so) and the working directory. With RAxML, the working directory specification requires the absolute path.

5. Let's inspect the outcome of the Biopython run. While the RAxML output is the same (save for stochasticity) for DendroPy and Biopython, DendroPy abstracts away a few things. With Biopython, you need to take care of the results yourself. You can also perform this with DendroPy, but in this case, it is optional:

```
from Bio import Phylo
biopython_tree = \
Phylo.read('bp_rx/RAxML_bestTree.biopython', 'newick')
```

   ❑ The preceding code will read the best tree from the RAxML run. The name of the file was appended with the project name that you specified in the previous step (in this case, Biopython).

   ❑ Take a look at the content of the bp_rx directory; here, you will find all the outputs from RAxML, including all 10 alternative trees.

Although the purpose of this book is not to teach phylogentic analysis, it's important to know why we do not inspect consensus and support information on the tree topology. You should research this in your dataset. For more information, refer to `http://www.geol.umd.edu/~tholtz/G331/lectures/cladistics5.pdf`.

Playing recursively with trees

This is not a book about programming on Python as the topic is vast. Having said that, it's not common for introductory Python books to discuss recursive programming at length. Recursive programming techniques are usually well tailored to deal with trees. They are also a common programming strategy with functional programming dialects, which can be quite useful when you perform concurrent development, which you can perform when processing very large datasets.

The phylogenetic notion of a tree is slightly different from computer science. Phylogenetic trees can be rooted (then they are normal tree data structures) or unrooted, making them undirected acyclic graphs. Phylogenetic trees can also have weights in their edges. Thus, be mindful of this when you read the documentation; if text is written by a phylogeneticist, you can expect the tree (rooted and unrooted), while most other documents will use undirected acyclic graphs for unrooted trees. Having said that, in this recipe, will assume that all trees are rooted.

Finally, note that while this recipe is devised mostly to help you understand recursive algorithms and tree-like structures, the final part is actually quite practical and fundamental for the next recipe to work.

Getting ready

You will need to have files from the previous recipe. As usual, you can find this content in the `05_Phylo/Trees.ipynb` notebook. We will use DendroPy's tree representations here. Note that most of this code is easily generalizable compared with other tree representations and libraries (phylogenetic or not).

How to do it...

Take a look at the following steps:

1. First, let's load the RAxML-generated tree for all *Ebola virus as follows*:

```
import dendropy
ebola_raxml = dendropy.Tree.get_from_path('my_ebola.nex',
    'nexus')
```

2. Then, compute the level of each node (the distance to the root node):

```
def compute_level(node, level=0):
    for child in node.child_nodes():
        compute_level(child, level + 1)
    if node.taxon is not None:
        print("%s: %d %d" % (node.taxon, node.level(),
            level))

compute_level(ebola_raxml.seed_node)
```

- DendroPy's node representation has a level method (which is used for comparison), but the point here is to introduce a recursive algorithm, so we will implement it anyway.

- Note how the function works; it's called with the `seed_node` (which is the root node under our assumption that we are dealing with rooted trees). The default level for the root node is 0. The function will then call itself for all its children nodes, increasing the level by one. Then, for each node that is not a leaf (it's internal to the tree), the calling will be repeated, and this will recurse until we get to the leaf nodes.

- For the leaf nodes, we then print the level (we could have done the same for the internal nodes) and show the same information computed by DendroPy's internal function.

3. Let's now compute the height of each node. The height of the node is the number of edges of the maximum downward path (going to the leaves) starting on that node as follows:

```
def compute_height(node):
    children = node.child_nodes()
    if len(children) == 0:
        height = 0
    else:
        height = 1 + max(map(lambda x: compute_height(x),
            children))
    desc = node.taxon or 'Internal'
    print("%s: %d %d" % (desc, height, node.level()))
    return height

compute_height(ebola_raxml.seed_node)
```

- ❑ Here, we will use the same recursive strategy, but each node will return its height to its parent; if the node is a leaf, then the height is 0; if not, then it's 1 plus the maximum of the height of its entire offspring.

- ❑ Note that we use a map over a lambda function to get all the heights of all the children of the current node. We then choose the maximum (the maximum function performs a reduce operation here because it summarizes all the values reported). If you are relating this to MapReduce frameworks, you are correct; these are inspired in functional programming dialects like these.

4. Let's now compute the number of offspring for each node; this should be quite easy to understand now:

```
def compute_nofs(node):
    children = node.child_nodes()
    nofs = len(children)
    map(lambda x: compute_nofs(x), children)
    desc = node.taxon or 'Internal'
    print("%s: %d %d" % (desc, nofs, node.level()))

compute_nofs(ebola_raxml.seed_node)
```

5. We will now print all the leaves (this is apparently trivial):

```
def print_nodes(node):
    for child in node.child_nodes():
        print_nodes(child)
    if node.taxon is not None:
        print('%s (%d)' % (node.taxon, node.level()))

print_nodes(ebola_raxml.seed_node)
```

- ❑ Note that all the functions that we have developed until now impose a very clear traversal pattern on the tree. You call your first offspring, then your offspring will call their offspring, and so on; only after this, you will call your next offspring in a depth-first pattern, but we can do things differently.

6. Let's now print the leaf nodes in a breath-first manner, that is, we will print first the leafs with the lowest level (closer to the root) as follows:

```
from collections import deque

def print_breadth(tree):
    queue = deque()
    queue.append(tree.seed_node)
    while len(queue) > 0:
        process_node = queue.popleft()
        if process_node.taxon is not None:
```

```
        print('%s (%d)' % (process_node.taxon,
            process_node.level()))
    else:
        for child in process_node.child_nodes():
            queue.append(child)
```

```
print_breadth(ebola_raxml)
```

- ❏ Before we explain the algorithm, let's see how different the result from this run will be compared to the previous one. For starters, take a look at the next figure. If you print the nodes by depth-first order, you will get **Y, A, X, B**, and **C**, but if you perform a breath-first traversal, you will get **X, B, C, Y**, and **A**. Tree traversal will have an impact on how the nodes are visited; more often than not that, this is important.

- ❏ Regarding the preceding code here, we will use a completely different approach as we will perform an iterative algorithm. We will use a **First-In First-Out** (**FIFO**) queue to help order our nodes. Note that Python's deque can be used efficiently as FIFO and also as **Last-In First-Out** (**LIFO**) because it implements an efficient data structure when you operate at both extremes.

- ❏ The algorithm starts by putting the root node on the queue. While the queue is not empty, we will take the node out front. If it's an internal node, we will put all its children on the queue.

- ❏ We will iterate the preceding step until the queue is empty. I encourage you to take a pen and paper and see how this works by performing this example on the figure by yourself because the code is small, but not trivial:

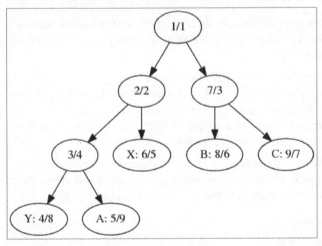

Figure 2: An example tree; the first number of each node indicates the order in which that node is visited with a depth-first algorithm; the second number indicates the order with a breadth-first algorithm

7. Let's get back to the real dataset. As we have a bit too much data to visualize, we will generate a trimmed down version, where we remove the subtrees that have single species (in the case of EBOV, have the same outbreak). We will also ladderize the tree, that is, sort the child nodes in order of the number of children:

```python
from copy import deepcopy
simple_ebola = deepcopy(ebola_raxml)

def simplify_tree(node):
    prefs = set()
    for leaf in node.leaf_nodes():
        my_toks = leaf.taxon.label.split(' ')
        if my_toks[0] == 'EBOV':
            prefs.add('EBOV' + my_toks[1])
        else:
            prefs.add(my_toks[0])
    if len(prefs) == 1:
        print(prefs, len(node.leaf_nodes()))
        node.taxon = dendropy.Taxon(label=list(prefs)[0])
        node.set_child_nodes([])
    else:
        for child in node.child_nodes():
            simplify_tree(child)

simplify_tree(simple_ebola.seed_node)
simple_ebola.ladderize()
simple_ebola.write_to_path('ebola_simple.nex', 'nexus')
```

❑ We will perform a deep copy of the tree structure. As our function and the ladderization is destructive (it will change the tree), we will want to maintain the original tree.

❑ DendroPy is able to enumerate all the leaf nodes (at this stage, a good exercise will be to write a function to perform this). With this functionality, we will get all leaves for a certain node. If they share the same species and the outbreak year in the case of EBOV, we remove all the child nodes, leaves, and internal subtree nodes.

❑ If they do not share the same species, we recurse down until that happens. The worst case is that when you are already at a leaf node, the algorithm trivially resolves to the species of the current node.

There's more...

There is a massive amount of computer science literature on the topic of trees and data structures; if you want to read more, Wikipedia provides a great introduction at `http://en.wikipedia.org/wiki/Tree_%28data_structure%29`.

Note that the use of lambda functions and map is not encouraged as a Python dialect; you can read some (old) opinion on the subject from Guido van Rossum at `http://www.artima.com/weblogs/viewpost.jsp?thread=98196`. I presented it here because it's a very common dialect with functional and recursive programming. The more common dialect will be based on list comprehensions (refer to `http://www.diveintopython3.net/comprehensions.html`).

In any case, the functional dialect based on using map and reduce is the conceptual base for MapReduce frameworks, and you can use frameworks such as Hadoop, Disco, or Spark to perform high-performance bioinformatics computing.

Visualizing phylogenetic data

Here, we will discuss how to visualize Phylogenetic trees. DendroPy has only simple visualization mechanisms based on drawing textual ASCII trees, but Biopython has quite a rich infrastructure, which we will leverage here.

Getting ready

This will require all the previous recipes. Remember that we will have files for the whole genus *Ebola virus*, including the RAxML tree. Furthermore, there will be a simplified genus version produced in the previous recipe.

Biopython uses matplotlib and graphviz as alternative backends. Graphviz is a graph visualization tool (do not confuse graph, the mathematical construct with a chart). While you should have matplotlib installed, you will need to install graphviz (`http://www.graphviz.org/`) and pygraphviz (`http://pygraphviz.github.io/`). Graphviz is available on Ubuntu and Linux (package graphviz). Pygraphviz can be installed using pip.

As usual, you can find this content in the `05_Phylo/Visualization.ipynb` notebook.

How to do it...

Take a look at the following steps:

1. Let's load all the phylogenetic data:

```
from copy import deepcopy
from Bio import Phylo
ebola_tree = Phylo.read('my_ebola.nex', 'nexus')
ebola_tree.name = 'Ebolavirus tree'
ebola_simple_tree = Phylo.read('ebola_simple.nex', 'nexus')
ebola_simple_tree.name = 'Ebolavirus simplified tree'
```

❏ For all trees that we read, we will change the name of the tree as the name will be printed later.

2. We can draw ASCII representations of the trees:

```
Phylo.draw_ascii(ebola_simple_tree)
Phylo.draw_ascii(ebola_tree)
```

❏ The ASCII representation of the simplified genus tree is shown in *Figure 3*. Here, we will not print the complete version because it will take several pages, but if you run the preceding code, you will be able to see that it's actually quite readable:

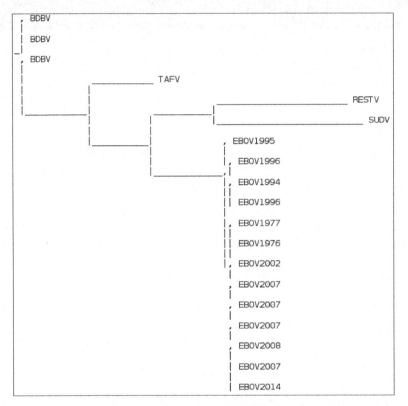

Figure 3: ASCII representation of a simplified Ebola virus dataset

3. **Bio.Phylo** allows the graphical representation of trees using matplotlib as a backend:

```
import matplotlib.pyplot as plt
fig = plt.figure(figsize=(16, 22))
ax = fig.add_subplot(111)
Phylo.draw(ebola_simple_tree, branch_labels=lambda c:
    c.branch_length if c.branch_length > 0.02 else None,
    axes=ax)
```

❑   In this case, we will print the branch lengths on the edges, but we will remove all lengths that are less than 0.02 to avoid clutter. The result is shown in the following figure:

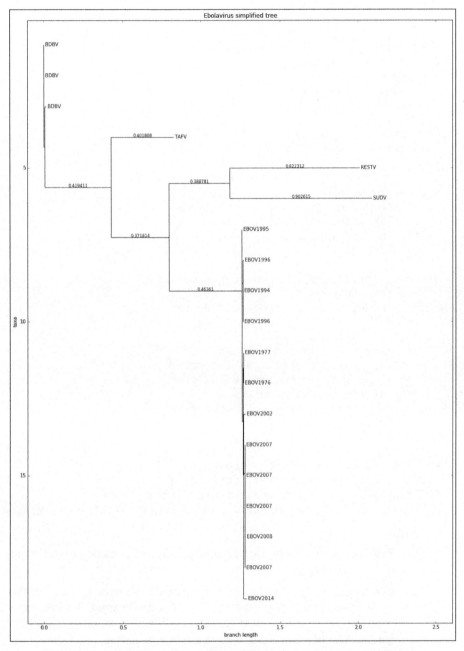

Figure 4: A matplotlib-based version of the simplified dataset with branch-lengths added

4. We will now plot the complete dataset, but we will color each bit of the tree differently. If a subtree has only a single virus species, it will get its own color. EBOV will have two colors; one for the 2014 outbreak and one for others as follows:

```
fig = plt.figure(figsize=(16, 22))
ax = fig.add_subplot(111)

from collections import OrderedDict
my_colors = OrderedDict({
    'EBOV_2014': 'red',
    'EBOV': 'magenta',
    'BDBV': 'cyan',
    'SUDV': 'blue',
    'RESTV' : 'green',
    'TAFV' : 'yellow'
})

def get_color(name):
    for pref, color in my_colors.items():
        if name.find(pref) > -1:
            return color
    return 'grey'

def color_tree(node, fun_color=get_color):
    if node.is_terminal():
        node.color = fun_color(node.name)
    else:
        my_children = set()
        for child in node.clades:
            color_tree(child, fun_color)
            my_children.add(child.color.to_hex())
        if len(my_children) == 1:
            node.color = child.color
        else:
            node.color = 'grey'

ebola_color_tree = deepcopy(ebola_tree)
color_tree(ebola_color_tree.root)
Phylo.draw(ebola_color_tree, axes=ax, label_func=
                  lambda x: x.name.split(' ')[0][1:] if x.name is
                  not None else None)
```

❏ This is a tree traversing algorithm, not unlike the ones presented in the previous recipe.

❏ As a recursive algorithm, it works as follows. If the node is a leaf, it will get a color based on its species (or the EBOV outbreak year). If it's an internal node and all the descendant nodes are of the same species below it, it will get the color of that specie; if there are several species after that, it will be colored in gray. Actually, the color function can be changed and will so later.

- ❑ Only the edge colors will be used (the labels will be printed in black).

- ❑ Note that Ladderization (performed in the previous recipe with DendroPy) helps quite a lot in terms of a clear visual appearance.

- ❑ We also deep copy the genus tree in order to color a copy; remember from the previous recipe that some tree traversal functions can change the state, and in this case, we want to preserve a version without any coloring.

- ❑ Note the usage of a lambda function to clean up the name that was changed by TrimAl, as shown in the following figure:

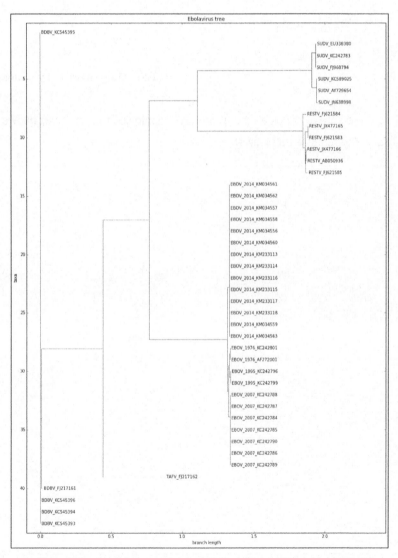

Figure 5: A colored phylogenetic tree with the complete Ebola virus dataset

5. Finally, Biopython's Phylo module allows the usage of graphviz, which is able to render graphs in an elegant way. We will use it with all the data as follows:

```
fig = plt.figure(figsize=(22, 22))
ax = fig.add_subplot(111)

def simplify_name(n):
    if n.is_terminal():
        return n.name[n.name.rfind('_') + 1:
                      n.name.find(' ')]
    else:
        return None

Phylo.draw_graphviz(ebola_color_tree,
                    label_func=simplify_name,
                    axes=ax, with_labels=True)
```

- Be sure to check all the alternative topologies that graphviz allows you to use
- The result is as follows:

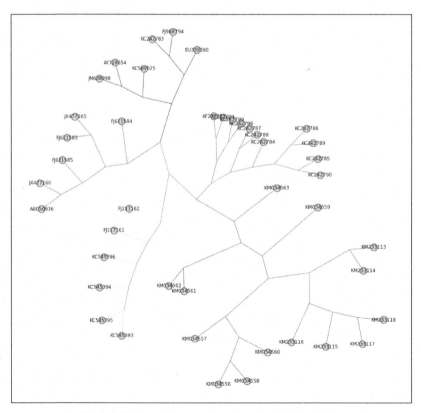

Figure 6: A colored-phylogenetic tree rendered with graphviz

There's more...

Tree and graph visualization is a complex topic; we will revisit this subject in *Chapter 8, Other Topics in Bioinformatics*, when we interface with Cytoscape. One interesting free software alternative (Java-based) for graph visualization is Gephi (`http://gephi.github.io/`). If you want to know more about the algorithms rendering trees and graphs, check the wikipedia page at `http://en.wikipedia.org/wiki/Graph_drawing` for an introduction to this fascinating topic.

7

Using the Protein Data Bank

In this chapter, we will cover the following recipes:

- ▶ Finding a protein in multiple databases
- ▶ Introducing Bio.PDB
- ▶ Extracting more information from a PDB file
- ▶ Computing molecular distances on a PDB file
- ▶ Performing geometric operations
- ▶ Implementing a basic PDB parser
- ▶ Animating with PyMol
- ▶ Parsing mmCIF files with Biopython

Introduction

Proteomics is the study of proteins that includes the protein function and structure. One of the main objectives of this field is to characterize the 3D structure of proteins. One of the most widely known computational resource in the Proteomics field is the Protein Data Bank, a repository with structural data of large biomolecules. Of course, there are also many databases that focus instead on protein primary structure; these are somewhat similar to genomic databases that we have seen in *Chapter 2, Next-generation Sequencing*.

In this chapter, we will mostly focus on processing data from the PDB. We will see how to parse PDB files, perform some geometric computations, and visualize molecules. We will use the old PDB file format because conceptually, it allows you to perform most necessary operations in a stable environment. Having said that, the newer mmCIF—slated to replace the PDB format—will also be presented in a later recipe. We will use Biopython and introduce PyMol for visualization. We will not discuss molecular docking here because this is probably more suitable a chemoinformatics book.

Throughout this chapter, we will use a classic example of a protein: Tumor protein p53, a protein involved in the regulation of the cell cycle (for example, apoptosis). This protein is highly related to cancer. There is plenty of information available about this protein on the Web.

Let's start with something that you should be more familiar with right now: accessing databases, especially for the protein primary structure (sequences of amino acids).

Finding a protein in multiple databases

Before we start performing some more structural biology, we will see how to access existing proteomic databases such as UniProt. We will query UniProt for our gene of interest: *TP53* and take it from there.

Getting ready

To access data, we will use Biopython and the REST API (we used a similar approach in *Chapter 3*, *Working with Genomes*) with the requests library to access web APIs. The requests API is an easy-to-use wrapper for web requests that can be installed using standard Python mechanisms (for example, pip and conda).

You can find this content in the `06_Prot/Intro.ipynb` notebook.

How to do it...

Take a look at the following steps:

1. First, let's define a function to perform REST queries on UniProt as follows:

```
import requests
server = 'http://www.uniprot.org/uniprot'
def do_request(server, ID='', **kwargs):
    params = ''
    req = requests.get('%s/%s%s' % (server, ID,
        params),params=kwargs)
```

```
if not req.ok:
    req.raise_for_status()
return req
```

❏ This is a simple function to perform REST queries. You may ask why are we not using the Biopython interface for this? Well, the current version of Biopython still refers to old ExPASy URLs, which do not work anymore because they have changed to UniProt. So, at this moment, this functionality is broken in Biopython 1.64. This is the consequence of Bioinformatics being a fast-moving, nonstable field, where sometimes software libraries do not keep up with changes in services.

2. We can now query all p53 genes that have been reviewed:

```
req = do_request(server, query='gene:p53 AND reviewed:yes',
                 format='tab',
                 columns='id,entry
name,length,organism,organism-id,database(PDB),database(HGNC)',
                 limit='50')
```

❏ We will query the p53 gene and request all entries that are reviewed (manually curated). The output will be in a tabular format. Probably, this is easiest to process. We request a maximum of 50 results, specifying the desired columns (for the complete list, refer to the following links).

❏ We could have restricted the output to just human data, but for this example, let's include all available species.

3. Let's check the result as follows:

```
import pandas as pd
import StringIO

uniprot_list = pd.read_table(StringIO.StringIO(req.text))
uniprot_list.rename(columns={'Organism ID': 'ID'},
        inplace=True)
print(uniprot_list)  # or just uniprot_list on IPython
```

- We will use pandas for easy processing of a tab-delimited list and pretty printing. The output of the IPython Notebook is as follows:

	Entry	Entry name	Length	Organism	ID	Cross-reference (PDB)	Cross-reference (HGNC)
0	Q9W678	P53_BARBU	369	Barbus barbus (Barbel) (Cyprinus barbus)	40830	NaN	NaN
1	Q29537	P53_CANFA	381	Canis familiaris (Dog) (Canis lupus familiaris)	9615	NaN	NaN
2	O09185	P53_CRIGR	393	Cricetulus griseus (Chinese hamster) (Cricetul...	10029	NaN	NaN
3	Q8SPZ3	P53_DELLE	387	Delphinapterus leucas (Beluga whale)	9749	NaN	NaN
4	P79892	P53_HORSE	280	Equus caballus (Horse)	9796	NaN	NaN
5	P04637	P53_HUMAN	393	Homo sapiens (Human)	9606	1A1U;1AIE;1C26;1DT7;1GZH;1H26;1HS5;1JSP;1KZY;1...	11998;
6	O93379	P53_ICTPU	376	Ictalurus punctatus (Channel catfish) (Silurus...	7998	NaN	NaN
7	P56423	P53_MACFA	393	Macaca fascicularis (Crab-eating macaque) (Cyn...	9541	NaN	NaN
8	P61260	P53_MACFU	393	Macaca fuscata fuscata (Japanese macaque)	9543	NaN	NaN
9	P56424	P53_MACMU	393	Macaca mulatta (Rhesus macaque)	9544	NaN	NaN
10	P02340	P53_MOUSE	387	Mus musculus (Mouse)	10090	1HU8;2GEQ;2IOI;2IOM;2IOO;2P52;3EXJ;3EXL;	NaN
11	P25035	P53_ONCMY	396	Oncorhynchus mykiss (Rainbow trout) (Salmo gai...	8022	NaN	NaN

4. Now, we can get the human p53 ID and use Biopython to retrieve and parse the `SwissProt` record:

```
from Bio import ExPASy, SwissProt
p53_human = uniprot_list[uniprot_list.ID ==
    9606]['Entry'].tolist()[0]
handle = ExPASy.get_sprot_raw(p53_human)
sp_rec= SwissProt.read(handle)
```

- Note that at this time, Biopython is up to date in terms of URLs to fetch a record (the `ExPASy.get_sprot_raw` call).

- We then use Biopython's `SwissProt` module to parse the record.

- 9606 is the NCBI taxonomic code for humans.

- As usual, if you get an error with network services, it may be a network or server problem. If this is the case, just retry at a later date.

5. Let's take a look at the p53 record as follows:

```
print(sp_rec.entry_name, sp_rec.sequence_length,
    sp_rec.gene_name)
print(sp_rec.description)
print(sp_rec.organism, sp_rec.seqinfo)
print(sp_rec.sequence)
```

```
('P53_HUMAN', 393, 'Name=TP53; Synonyms=P53;')
RecName: Full=Cellular tumor antigen p53; AltName: Full=Antigen NY-CO-13; AltName: Full=Phosphoprotein p53; AltName: Full=Tumor sup
pressor p53;
('Homo sapiens (Human).', (393, 43653, 'AD5C149FD8106131'))
MEEPQSDPSVEPPLSQETFSDLWKLLPENNVLSPLPSQAMDDLMLSPDDIEQWFTEDPGPDEAPRMPEAAPPVAPAPAAPTPAAPAPAPSWPLSSSVPSQKTYQGSYGFRLGFLHSGTAKSVTCTYSPALN
KMFCQLAKTCPVQLWVDSTPPPGTRVRAMAIYKQSQHMTEVVRRCPHHERCSDSDGLAPPQHLIRVEGNLRVEYLDDRNTFRHSVVVPYEPPEVGSDCTTIHYNYMCNSSCMGGMNRRPILTIITLEDSSG
NLLGRNSFEVRVCACPGRDRRTEEENLRKKGEPHHELPPGSTKRALPNNTSSSPQPKKKPLDGEYFTLQIRGRERFEMFRELNEALELKDAQAGKEPGGSRAHSSHLKSKKGQSTSRHKKLMFKTEGPDSD
```

6. A deeper look at the preceding record reveals a lot of really interesting information, especially on features, **Gene Ontology** (**GO**), and database cross-references:

```
from collections import defaultdict
done_features = set()
print(len(sp_rec.features))
for feature in sp_rec.features:
    if feature[0] in done_features:
        continue
    else:
        done_features.add(feature[0])
        print(feature)
print(len(sp_rec.cross_references))
per_source = defaultdict(list)
for xref in sp_rec.cross_references:
    source = xref[0]
    per_source[source].append(xref[1:])
print(per_source.keys())
done_GOs = set()
print(len(per_source['GO']))
for annot in per_source['GO']:
    if annot[1][0] in done_GOs:
        continue
    else:
        done_GOs.add(annot[1][0])
        print(annot)
```

Note that we are not even printing the whole information here, just a summary of it. We print a number of features on the sequence with one example per type, a number of external database references plus databases that are referred, and a number of GO entries along with three examples. Currently, there are 1493 features, 604 external references, and 133 GO terms just for this protein:

```
1493
('CHAIN', 1, 393, 'Cellular tumor antigen p53.', 'PRO_0000185703')
('DNA_BIND', 102, 292, '', '')
('REGION', 1, 83, 'Interaction with HRMT1L2.', '')
('MOTIF', 17, 25, 'TAD1.', '')
('METAL', 176, 176, 'Zinc.', '')
('SITE', 120, 120, 'Interaction with DNA.', '')
('MOD_RES', 9, 9, 'Phosphoserine; by HIPK4. {ECO:0000269|PubMed:18022393}.', '')
('CROSSLNK', 291, 291, 'Glycyl lysine isopeptide (Lys-Gly) (interchain with G-Cter in ubiquitin). {ECO:0000269|PubMed:19536131}.',
'')
('VAR_SEQ', 1, 132, 'Missing (in isoform 7, isoform 8 and isoform 9). {ECO:0000303|PubMed:16131611}.', 'VSP_040833')
('VARIANT', 5, 5, 'Q -> H (in a sporadic cancer; somatic mutation; abolishes strongly phosphorylation).', 'VAR_044543')
('MUTAGEN', 15, 15, 'S->A: Loss of interaction with PPP2R5C, PPP2CA AND PPP2R1A. {ECO:0000269|PubMed:17967874}.', '')
('HELIX', 19, 23, '{ECO:0000244|PDB:3DAC}.', '')
('STRAND', 27, 29, '{ECO:0000244|PDB:2K8F}.', '')
('TURN', 105, 108, '{ECO:0000244|PDB:3D06}.', '')
604
['GeneReviews', 'DNASU', 'MIM', 'SUPFAM', 'Genevestigator', 'HOVERGEN', 'ExpressionAtlas', 'MaxQB', 'GeneWiki', 'SMR', 'Orphanet',
'CTD', 'GO', 'PhylomeDB', 'CCDS', 'neXtProt', 'BindingDB', 'RefSeq', 'PRIDE', 'DMDM', 'Reactome', 'PROSITE', 'TreeFam', 'SWISS-2DPA
GE', 'NextBio', 'DIP', 'PRO', 'PANTHER', 'TCDB', 'Gene3D', 'DrugBank', 'PMAP-CutDB', 'Bgee', 'EvolutionaryTrace', 'ChEMBL', 'PIR',
'InParanoid', 'GeneCards', 'Pfam', 'PDBsum', 'KEGG', 'eggNOG', 'EMBL', 'PaxDb', 'DisProt', 'Proteomes', 'ProteinModelPortal', 'Ense
mbl', 'ChiTaRS', 'SignaLink', 'HPA', 'IntAct', 'MINT', 'PDB', 'UniGene', 'OMA', 'InterPro', 'PharmGKB', 'PhosphoSite', 'GenomeRNAi'
, 'KO', 'BioGrid', 'UCSC', 'HGNC', 'PRINTS', 'GeneTree', 'GeneID']
133
('GO:0000785', 'C:chromatin', 'IBA:GO_Central')
('GO:0005524', 'F:ATP binding', 'IDA:UniProtKB')
('GO:0006915', 'P:apoptotic process', 'TAS:Reactome')
```

There's more...

There are many more databases with information on proteins, some of these are referred in the preceding record; you can explore its result to try to find data elsewhere.

For detailed information about UniProt's REST interface, refer to `http://www.uniprot.org/help/programmatic_access`.

Introducing Bio.PDB

Here, we will introduce Biopython's PDB module to deal with the Protein Data Bank. We will use three models that represent part of the p53 protein. You can read more about these files and p53 at `http://www.rcsb.org/pdb/101/motm.do?momID=31`.

Getting ready

You should be aware of the basic PDB data model of `Model/Chain/Residue/Atom` objects. A good explanation on Biopython's Structural Bioinformatics FAQ can be found at `http://biopython.org/wiki/The_Biopython_Structural_Bioinformatics_FAQ`.

You can find this content in the `06_Prot/PDB.ipynb` notebook.

Of the three models that we will download, the `1TUP` model will be used in the remaining recipes. Take some time to study this model as it will help you later on.

How to do it...

Take a look at the following steps:

1. First, let's retrieve our models of interest as follows:

```
from __future__ import print_function
from Bio import PDB
repository = PDB.PDBList()
repository.retrieve_pdb_file('1TUP', pdir='.')
repository.retrieve_pdb_file('1OLG', pdir='.')
repository.retrieve_pdb_file('1YCQ', pdir='.')
```

   Note that Bio.PDB can take care of downloading files for you. Moreover, these download will only occur if no local copy is already present.

2. Let's parse our records, as shown in the following code:

```
parser = PDB.PDBParser()
p53_1tup = parser.get_structure('P 53 - DNA Binding',
'pdb1tup.ent')
p53_1olg = parser.get_structure('P 53 - Tetramerization',
'pdb1olg.ent')
p53_1ycq = parser.get_structure('P 53 - Transactivation',
'pdb1ycq.ent')
```

   You may get some warnings about the content of the file. These are usually not problematic.

3. Let's inspect our headers as follows:

```
def print_pdb_headers(headers, indent=0):
    ind_text = ' ' * indent
    for header, content in headers.items():
        if type(content) == dict:
            print('\n%s%20s:' % (ind_text, header))
            print_pdb_headers(content, indent + 4)
            print()
        elif type(content) == list:
            print('%s%20s:' % (ind_text, header))
            for elem in content:
                print('%s%21s %s' % (ind_text, '->', elem))
        else:
            print('%s%20s: %s' % (ind_text, header,
content))

print_pdb_headers(p53_1tup.header)
```

Headers are parsed as a dictionary of dictionaries. As such, we will use a recursive function to parse them. This function will increase indentation for ease of reading, and annotate lists of elements with the → prefix. For more details on recursive functions, refer to the previous chapter. Part of the output is as follows:

```
  structure_method: x-ray diffraction
           head: antitumor protein/dna
        journal: AUTH   Y.CHO,S.GORINA,P.D.JEFFREY,N.P.PAVLETICHTITL   CRYSTAL STRUCTURE OF A P53 TUMOR SUPPRESSOR-DNATITL 2 C
OMPLEX: UNDERSTANDING TUMORIGENIC MUTATIONS.REF     SCIENCE                 V. 265   346 1994REFN                      ISSN 0036-
8075PMID   8023157
   journal_reference: y.cho,s.gorina,p.d.jeffrey,n.p.pavletich crystal structure of a p53 tumor suppressor-dna complex: understandi
ng tumorigenic mutations. science v. 265 346 1994 issn 0036-8075 8023157

        compound:

            1:
              molecule: dna (5'-d(*tp*tp*tp*cp*cp*tp*ap*gp*ap*cp*tp*tp*gp*cp*cp*a p*ap*tp*tp*a)-3')
                misc:
            engineered: yes
                chain: e


            3:
              molecule: protein (p53 tumor suppressor )
                misc:
            engineered: yes
                chain: a, b, c


            2:
              molecule: dna (5'-d(*ap*tp*ap*ap*tp*tp*gp*gp*gp*cp*ap*ap*gp*tp*cp*tp*a p*gp*gp*ap*a)-3')
                misc:
            engineered: yes
                chain: f


        keywords: antigen p53, antitumor protein/dna complex
            name:  tumor suppressor p53 complexed with dna
          author: Y.Cho,S.Gorina,P.D.Jeffrey,N.P.Pavletich
  deposition_date: 1995-07-11
     release_date: 1995-07-11
```

4. We want to know the content of each chain on these files; for this, let's take a look at the COMPND records:

```
print(p53_1tup.header['compound'])
print(p53_1olg.header['compound'])
print(p53_1ycq.header['compound'])
```

❑   This will return all compound headers printed in the preceding code. Unfortunately, this is not the best way to get information on chains. An alternative will be to get DBREF records, but Biopython's parser is currently not able to access these. We will have a recipe to deal with this, but for now, this is what the parser can do.

❑   Having said that, using a tool like **grep** will easily extract this information.

❑   Note that for 1TUP, chains A, B, and C chains are from the protein, and E and F chains are from the DNA. This information will be useful in the future.

5. Let's do a top-down analysis of each PDB file. For now, let's just get all chains, the number of residues, and atoms per chain as follows:

```python
def describe_model(name, pdb):
    print()
    for model in pdb:
        for chain in model:
            print('%s - Chain: %s. Number of residues: %d.
Number of atoms: %d.' %
                  (name, chain.id, len(chain),
                   len(list(chain.get_atoms()))))

describe_model('1TUP', p53_1tup)
describe_model('1OLG', p53_1olg)
describe_model('1YCQ', p53_1ycq)
```

> ❏ We will perform a bottom-up approach in a later recipe. Here is the output for 1TUP:

```
1TUP - Chain: E. Number of residues: 43. Number of atoms: 442.
1TUP - Chain: F. Number of residues: 35. Number of atoms: 449.
1TUP - Chain: A. Number of residues: 395. Number of atoms: 1734.
1TUP - Chain: B. Number of residues: 265. Number of atoms: 1593.
1TUP - Chain: C. Number of residues: 276. Number of atoms: 1610.
```

6. Let's get all nonstandard residues (HETATM) with the exception of water in the 1TUP model, as shown in the following code:

```python
for residue in p53_1tup.get_residues():
    if residue.id[0] in [' ', 'W']:
        continue
    print(residue.id)
```

> ❏ We have three zincs, one on each of the protein chains.

7. Let's take a look at a residue:

```python
res = next(p53_1tup[0]['A'].get_residues())
print(res)
for atom in res:
    print(atom, atom.serial_number, atom.element)
p53_1tup[0]['A'][94]['CA']
```

❑ This will print all atoms on a certain residue:

```
<Residue SER het=  resseq=94 icode= >
<Atom N>  858 N
<Atom CA>  859 C
<Atom C>  860 C
<Atom O>  861 O
<Atom CB>  862 C
<Atom OG>  863 O
<Atom CA>
```

❑ Note the last statement. It is there just to show that you can directly access an atom by resolving model, chain, residue, and finally the atom.

8. Finally, let's export the protein fragment to a FASTA file as follows:

```
from Bio.SeqIO import PdbIO, FastaIO

def get_fasta(pdb_file, fasta_file, transfer_ids=None):
    fasta_writer = FastaIO.FastaWriter(fasta_file)
    fasta_writer.write_header()
    for rec in PdbIO.PdbSeqresIterator(pdb_file):
        if len(rec.seq) == 0:
            continue
        if transfer_ids is not None and rec.id not in \
            transfer_ids:
            continue
        print(rec.id, rec.seq, len(rec.seq))
        fasta_writer.write_record(rec)

get_fasta(open('pdb1tup.ent'), open('1tup.fasta', 'w'),
    transfer_ids=['1TUP:B'])
get_fasta(open('pdb1olg.ent'), open('1olg.fasta', 'w'),
    transfer_ids=['1OLG:B'])
get_fasta(open('pdb1ycq.ent'), open('1ycq.fasta', 'w'),
    transfer_ids=['1YCQ:B'])
```

❑ If you inspect the protein chain, you will see that they are equal in each model, so we export a single one. In the case of `1YCQ`, we export the smallest one because the biggest one is not p53-related.

❑ As you can see, here we are using Bio.SeqIO, not Bio.PDB.

There's more...

The PDB parser is incomplete. It's not very likely that a complete parser will be seen soon as the community migrates to the mmCIF format. However, if you need to parse a PDB file, refer to the parsing recipe in this chapter.

Although the future is the mmCIF format (`http://mmcif.wwpdb.org/`), PDB files are still around. Conceptually, many operations are similar after you have parsed the file.

If you are on Python 3, I suggest you take a look at PyProt at `https://github.com/rasbt/pyprot`.

Extracting more information from a PDB file

Here, we will continue our exploration of the record structure produced by Bio.PDB from PDB files.

Getting ready

For general information about the PDB models that we are using, refer to the previous recipe.

You can find this content in the `06_Prot/Stats.ipynb` notebook.

How to do it...

Take a look at the following steps:

1. First, let's retrieve `1TUP` as follows:

   ```
   from __future__ import print_function
   from Bio import PDB
   repository = PDB.PDBList()
   parser = PDB.PDBParser()
   repository.retrieve_pdb_file('1TUP', pdir='.')
   p53_1tup = parser.get_structure('P 53', 'pdb1tup.ent')
   ```

2. Then, extract some atom-related statistics:

   ```
   from collections import defaultdict
   atom_cnt = defaultdict(int)
   atom_chain = defaultdict(int)
   atom_res_types = defaultdict(int)

   for atom in p53_1tup.get_atoms():
       my_residue = atom.parent
       my_chain = my_residue.parent
       atom_chain[my_chain.id] += 1
       if my_residue.resname != 'HOH':
           atom_cnt[atom.element] += 1
       atom_res_types[my_residue.resname] += 1
   ```

```
print(dict(atom_res_types))
print(dict(atom_chain))
print(dict(atom_cnt))
```

❑ This will print information on the atom's residue type, number of atoms per chain, and quantity per element, as shown in the following screenshot:

```
{'ILE': 144, 'GLN': 189, ' ZN': 3, 'THR': 294, 'HOH': 384, 'GLY': 156, 'ASP': 192, 'PHE': 165, 'TRP': 42, 'GLU': 297, 'CYS': 180, '
HIS': 210, 'SER': 323, 'LYS': 135, ' DG': 176, 'PRO': 294, ' DC': 152, ' DA': 270, 'ALA': 105, 'MET': 144, 'LEU': 336, 'ARG': 561,
' DT': 257, 'VAL': 315, 'ASN': 216, 'TYR': 288}
{'A': 1734, 'C': 1610, 'B': 1593, 'E': 442, 'F': 449}
{'P': 40, 'ZN': 3, 'S': 48, 'C': 3238, 'O': 1114, 'N': 1001}
```

❑ Note that the preceding number of residues is not the proper number of residues, but the amount of times that a certain residue type is referred (it adds up to the number of atoms, not residues).

❑ Note the water (W), nucleotide (DA, DC, DG, and DT), and Zinc (ZN) residues which add to the amino acid ones.

3. Now, let's now count the instances per residue and the number of residues per chain:

```
res_types = defaultdict(int)
res_per_chain = defaultdict(int)
for residue in p53_1tup.get_residues():
    res_types[residue.resname] += 1
    res_per_chain[residue.parent.id] +=1
print(dict(res_types))
print(dict(res_per_chain))
```

```
{'ILE': 18, 'GLN': 21, ' ZN': 3, 'THR': 42, 'HOH': 384, 'GLY': 39, 'ASP': 24, 'PHE': 15, 'TRP': 3, 'GLU': 33, 'CYS': 30, 'HIS': 21,
'SER': 54, 'LYS': 15, ' DG': 8, 'PRO': 42, ' DC': 8, ' DA': 13, 'ALA': 21, 'MET': 18, 'LEU': 42, 'ARG': 51, ' DT': 13, 'VAL': 45,
'ASN': 27, 'TYR': 24}
{'A': 395, 'C': 276, 'B': 265, 'E': 43, 'F': 35}
```

4. We can also get the bounds of a set of atoms:

```
import sys
def get_bounds(my_atoms):
    my_min = [sys.maxint] * 3
    my_max = [-sys.maxint] * 3
    for atom in my_atoms:
        for i, coord in enumerate(atom.coord):
            if coord < my_min[i]:
                my_min[i] = coord
            if coord > my_max[i]:
                my_max[i] = coord
    return my_min, my_max
chain_bounds = {}
```

```
for chain in p53_1tup.get_chains():
    print(chain.id, get_bounds(chain.get_atoms()))
    chain_bounds[chain.id] = get_bounds(chain.get_atoms())
print(get_bounds(p53_1tup.get_atoms()))
```

- ❑ A set of atoms can be a whole model, a chain, a residue, or any subset you are interested in. In this case, we will print boundaries for all chains and the whole model. Numbers convey little intuition, so we get a little bit more graphical.

- ❑ `sys.maxint` does not exist on Python 3; here, you may use `sys.maxsize` instead.

5. To have a notion of the size of each chain, a plot is probably more informative than the numbers in the preceding code:

```
import matplotlib.pyplot as plt
from mpl_toolkits.mplot3d import Axes3D
fig = plt.figure(figsize=(16, 9))
ax3d = fig.add_subplot(111, projection='3d')
ax_xy = fig.add_subplot(331)
ax_xy.set_title('X/Y')
ax_xz = fig.add_subplot(334)
ax_xz.set_title('X/Z')
ax_zy = fig.add_subplot(337)
ax_zy.set_title('Z/Y')
color = {'A': 'r', 'B': 'g', 'C': 'b', 'E': '0.5', 'F':
    '0.75'}
zx, zy, zz = [], [], []
for chain in p53_1tup.get_chains():
    xs, ys, zs = [], [], []
    for residue in chain.get_residues():
        ref_atom = next(residue.get_iterator())
        x, y, z = ref_atom.coord
        if ref_atom.element == 'ZN':
            zx.append(x)
            zy.append(y)
            zz.append(z)
            continue
        xs.append(x)
        ys.append(y)
        zs.append(z)
    ax3d.scatter(xs, ys, zs, color=color[chain.id])
    ax_xy.scatter(xs, ys, marker='.',
        color=color[chain.id])
    ax_xz.scatter(xs, zs, marker='.',
        color=color[chain.id])
```

```
       ax_zy.scatter(zs, ys, marker='.',
           color=color[chain.id])
ax3d.set_xlabel('X')
ax3d.set_ylabel('Y')
ax3d.set_zlabel('Z')
ax3d.scatter(zx, zy, zz, color='k', marker='v', s=300)
ax_xy.scatter(zx, zy, color='k', marker='v', s=80)
ax_xz.scatter(zx, zz, color='k', marker='v', s=80)
ax_zy.scatter(zz, zy, color='k', marker='v', s=80)
for ax in [ax_xy, ax_xz, ax_zy]:
    ax.get_yaxis().set_visible(False)
    ax.get_xaxis().set_visible(False)
```

- There are plenty of molecular visualization tools. Indeed, we will discuss PyMol later. However, matplotlib is enough for some simple visualization. The most important point about matplotlib is that it's stable and very easy to integrate into a reliable production code.

- In the preceding chart, we performed a 3D plot of chains, the DNA in gray, and the protein chains in different colors.

- We also plot planar projections (X/Y, X/Z, and Z/Y) on the left-hand side in the following figure:

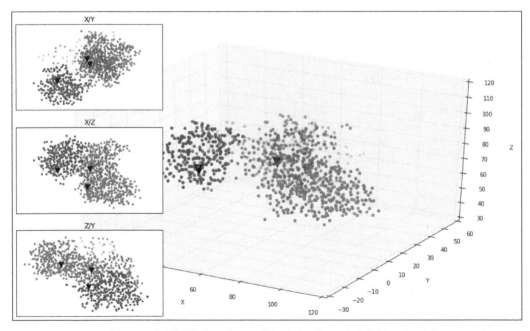

Figure 1: The spatial distribution of the protein chains; the main figure is a 3D plot and the left subplots are planar views (X/Y, X/Z, and Z/Y)

Computing molecular distances on a PDB file

Here, we will find atoms closer to three zincs on the 1TUP model. We will consider several distances to these zincs. We will take this opportunity to discuss the performance of algorithms.

Getting ready

You can find this content in the 06_Prot/Distance.ipynb notebook. Take a look at the Bio.PDB recipe.

How to do it...

Take a look at the following steps:

1. Let's load our model as follows:

```
from __future__ import print_function
from Bio import PDB
repository = PDB.PDBList()
parser = PDB.PDBParser()
repository.retrieve_pdb_file('1TUP', pdir='.')
p53_1tup = parser.get_structure('P 53', 'pdb1tup.ent')
```

2. We will now get our zincs against which we perform later comparisons:

```
zns = []
for atom in p53_1tup.get_atoms():
    if atom.element == 'ZN':
        zns.append(atom)
for zn in zns:
        print(zn, zn.coord)
```

   ❑ You should see three zinc atoms.

3. Then, let's define a function to get the distance between one atom and a set of other atoms as follows:

```
import math
def get_closest_atoms(pdb_struct, ref_atom, distance):
    atoms = {}
    rx, ry, rz = ref_atom.coord
    for atom in pdb_struct.get_atoms():
        if atom == ref_atom:
            continue
        x, y, z = atom.coord
```

```
        my_dist = math.sqrt((x - rx)**2 + (y - ry)**2 + (z
            - rz)**2)
        if my_dist < distance:
            atoms[atom] = my_dist
    return atoms
```

- ❑ We get coordinates for our reference atom and then iterate over our desired comparison list. If an atom is close enough, it's added to the return list.

4. Also, we now compute atoms near our zincs, which is up to 4 Ångström for our model:

```
for zn in zns:
    print()
    print(zn.coord)
    atoms = get_closest_atoms(p53_1tup, zn, 4)
    for atom, distance in atoms.items():
        print(atom.element, distance, atom.coord)
```

- ❑ Here, we show the result for the first zinc, including the element, distance, and coordinates, as shown in the following screenshot:

```
[ 58.10800171  23.24200058  57.42399979]
S 2.27892096249 [ 60.31700134  23.31800079  57.97900009]
C 2.92437196013 [ 56.99300003  23.94300079  54.81299973]
C 3.40801176963 [ 57.77000046  21.2140007   60.14199829]
S 2.32622437996 [ 57.06499863  21.45199966  58.48199844]
C 3.62725094648 [ 61.61000061  24.08699989  57.00099945]
C 3.85772919812 [ 61.14799881  25.06100082  55.89699936]
C 3.45665374923 [ 58.88600159  20.86700058  55.0359993 ]
C 3.08721447067 [ 57.20500183  25.09900093  59.71900177]
C 3.06412055976 [ 58.04700089  22.03800011  54.60699844]
S 2.22531584465 [ 56.91400146  25.05400085  57.91699982]
N 1.99182735373 [ 57.75500107  23.07299995  55.47100067]
```

- ❑ We only have three zincs, so the number of computations is quite reduced, but now imagine that we had more or that we were doing a pairwise comparison among all atoms in the set (remember that the number of comparisons grow quadratically with the number of atoms in a pairwise case). Although our case is small, it's not difficult to forecast use cases with much more comparisons taking a lot of time. We will get back to this soon.

5. Let's see how many atoms we get as we increase the distance:

```
for distance in [1, 2, 4, 8, 16, 32, 64, 128]:
    my_atoms = []
    for zn in zns:
        atoms = get_closest_atoms(p53_1tup, zn, distance)
        my_atoms.append(len(atoms))
    print(distance, my_atoms)
```

❏ The result is as follows:

```
1 [0, 0, 0]
2 [1, 0, 0]
4 [11, 11, 12]
8 [109, 113, 106]
16 [523, 721, 487]
32 [2381, 3493, 2053]
64 [5800, 5827, 5501]
128 [5827, 5827, 5827]
```

6. As we have seen before, this specific case is not very expensive, but let's time this anyway, as shown in the following code:

```
import timeit
nexecs = 10
print(timeit.timeit('get_closest_atoms(p53_1tup, zns[0],
4.0)',
                    'from __main__ import
get_closest_atoms, p53_1tup, zns',
                    number=nexecs) / nexecs * 1000)
```

❏ Here, we will use the `timeit` module to execute this function 10 times and then print the result in milliseconds. We pass the function as a string and pass yet another string with the necessary imports to make this function work. If you are on the IPython Notebook, you are probably aware of the `%timeit` magic and how it makes your life much easier in this case.

❏ This takes roughly 42 milliseconds on the machine where the code was tested. Obviously, on your computer, you will get somewhat different results.

7. Can we do better? Let's consider a different distance function, as shown in the following code:

```
def get_closest_alternative(pdb_struct, ref_atom,
    distance):
    atoms = {}
    rx, ry, rz = ref_atom.coord
    for atom in pdb_struct.get_atoms():
        if atom == ref_atom:
            continue
        x, y, z = atom.coord
        if abs(x - rx) > distance or abs(y - ry) > distance \
            or abs(z - rz) > distance:
            continue
        my_dist = math.sqrt((x - rx)**2 + (y - ry)**2 + (z
            - rz)**2)
        if my_dist < distance:
            atoms[atom] = my_dist
    return atoms
```

So, we take the original function and add a very simplistic `if` with distances. The rationale for this is that the computational cost of the square root and may be the float power operation is very expensive, so we will try to avoid it. However, for all atoms that are closer than the target distance in any dimension, this function will be more expensive.

8. Now, let's time against it:

```
print(timeit.timeit('get_closest_alternative(p53_1tup,
zns[0], 4.0)',
                    'from __main__ import
get_closest_alternative, p53_1tup, zns',
                    number=nexecs) / nexecs * 1000)
```

- On the same machine as the preceding example, we have **16** milliseconds, that is, roughly three times faster.

9. However, is it always better? Let's now compare the cost with different distances as follows:

```
print('Standard')
for distance in [1, 4, 16, 64, 128]:
    print(timeit.timeit('get_closest_atoms(p53_1tup,
zns[0], distance)',
                        'from __main__ import
get_closest_atoms, p53_1tup, zns, distance',
                        number=nexecs) / nexecs * 1000)
print('Optimized')
for distance in [1, 4, 16, 64, 128]:
    print(timeit.timeit('get_closest_alternative(p53_1tup,
zns[0], distance)',
                        'from __main__ import get_closest_
alternative, p53_1tup, zns, distance',
                        number=nexecs) / nexecs * 1000)
```

- The result is shown in the following screenshot:

```
Standard
45.9681987762
45.431804657
43.1920051575
43.0327892303
42.4966812134
Optimized
14.5443201065
15.860414505
27.8204917908
65.1834011078
65.1820898056
```

- ❏ Note that the cost of the standard version is mostly constant, whereas the optimized version varies with the distance to get the closest atoms; larger the distance, more cases will be computed using the extra if plus the square root, making the function more expensive.

- ❏ The larger point here is that you can probably code functions that are more efficient using smart computation shortcuts, but the complexity cost may change qualitatively. In the preceding case, I suggest that the second function is more efficient for all realistic and interesting cases when you try to find the closest atoms. However, you have to be careful while designing your own versions of optimized algorithms.

- ❏ We will revisit this subject in the very last chapter.

Performing geometric operations

We will now perform computations with geometry information, including computing the center of mass of chains and of whole models.

Getting ready

You can find this content in the `06_Prot/Mass.ipynb` notebook.

How to do it...

Take a look at the following steps:

1. First, let's retrieve the data:

```
from __future__ import print_function
import numpy as np
from Bio import PDB
repository = PDB.PDBList()
parser = PDB.PDBParser()
repository.retrieve_pdb_file('1TUP', pdir='.')
p53_1tup = parser.get_structure('P 53', 'pdb1tup.ent')
```

2. Then, remember the type of residues that we have with the following code:

```
my_residues = set()
for residue in p53_1tup.get_residues():
    my_residues.add(residue.id[0])
print(my_residues)
```

- ❏ So, we have H_ ZN (zinc) and W (water) that are HETATMs and the vast majority that are standard PDB ATOMs.

3. Let's compute the masses for all chains, zincs and waters using the following code:

```
def get_mass(atoms, accept_fun=lambda atom:
    atom.parent.id[0] != 'W'):
    return sum([atom.mass for atom in atoms if
        accept_fun(atom)])

chain_names = [chain.id for chain in p53_1tup.get_chains()]
my_mass = np.ndarray((len(chain_names), 3))
for i, chain in enumerate(p53_1tup.get_chains()):
    my_mass[i, 0] = get_mass(chain.get_atoms())
    my_mass[i, 1] = get_mass(chain.get_atoms(),
accept_fun=lambda atom: atom.parent.id[0] not in [' ',
'W'])
    my_mass[i, 2] = get_mass(chain.get_atoms(),
accept_fun=lambda atom: atom.parent.id[0] == 'W')
masses = pd.DataFrame(my_mass, index=chain_names,
    columns=['No Water','Zincs', 'Water'])
print(masses) # masses
```

- ❑ The `get_mass` function returns the mass of all atoms in the list that pass an acceptance criterion function. The default acceptance criterion is not to be a water residue.

- ❑ We then compute the mass for all chains. We have three versions: just amino acids, zincs, and waters. Zinc does nothing more than detecting a single atom per chain in this model. The output is as follows:

	No Water	Zincs	Water
E	6068.04412	0.00	351.9868
F	6258.20442	0.00	223.9916
A	20548.26300	65.39	3167.8812
B	20368.18840	65.39	1119.9580
C	20466.22540	65.39	1279.9520

4. Let's compute the geometric center and center of mass of the model as follows:

```
def get_center(atoms, weight_fun=lambda atom: 1 if
    atom.parent.id[0] != 'W' else 0):
    xsum = ysum = zsum = 0.0
    acum = 0.0
    for atom in atoms:
        x, y, z = atom.coord
        weight = weight_fun(atom)
        acum += weight
```

```
        xsum += weight * x
        ysum += weight * y
        zsum += weight * z
    return xsum / acum, ysum / acum, zsum / acum

print(get_center(p53_1tup.get_atoms()))
print(get_center(p53_1tup.get_atoms(),
                        weight_fun=lambda atom: atom.mass if
                        atom.parent.id[0] != 'W' else 0))
```

- ❏ First, we define a weighted function to get coordinates of the center. The default function will treat all atoms as equal as long as they are not a water residue.

- ❏ We then compute the geometric center and the center of mass by redefining the weight function with a value of each atom equal to its mass. The geometric center is computed irrespective of its molecular weights.

- ❏ For example, you may want to compute the center of mass of the protein without DNA chains.

5. Let's compute the center of mass and the geometric center of each chain as follows:

```
my_center = np.ndarray((len(chain_names), 6))
for i, chain in enumerate(p53_1tup.get_chains()):
    x, y, z = get_center(chain.get_atoms())
    my_center[i, 0] = x
    my_center[i, 1] = y
    my_center[i, 2] = z
    x, y, z = get_center(chain.get_atoms(), weight_fun=lambda
    atom: atom.mass if atom.parent.id[0] !=
    'W' else 0)
    my_center[i, 3] = x
    my_center[i, 4] = y
    my_center[i, 5] = z
weights = pd.DataFrame(my_center, index=chain_names,
    columns=['X', 'Y', 'Z', 'X (Mass)', 'Y (Mass)', 'Z
    (Mass)'])
print(weights) # weights
```

The result is as shown here:

	X	Y	Z	X (Mass)	Y (Mass)	Z (Mass)
E	49.727231	32.744879	81.253417	49.708513	32.759725	81.207395
F	51.982368	33.843370	81.578795	52.002223	33.820064	81.624394
A	72.990763	28.825429	56.714012	72.822668	28.810327	56.716117
B	67.810026	12.624435	88.656590	67.729100	12.724130	88.545659
C	38.221565	-5.010494	88.293141	38.169364	-4.915395	88.166711

There's more...

Although this is not a book based on the protein structure determination technique, remember that X-ray crystallography methods cannot detect hydrogens, so computing the mass of residues might be based on very inaccurate models; refer to http://www.umass.edu/microbio/chime/pe_beta/pe/protexpl/help_hyd.htm.

Implementing a basic PDB parser

As you know, by now the Bio.PDB parser is not complete. Here, we will develop a framework that allows you to parse other records on PDB files. Although we can expect a migration from PDB to the mmCIF format in the future, this is still useful in many situations.

Getting ready

In order to parse a format, we need its specification. You can find this at http://www.wwpdb.org/documentation/file-format.php. We will mostly be concerned with secondary structure records (HELIX and SHEET), but you will find more records in your scaffold parser. You can extend this scaffold to other records that you may need.

You can find this content in the 06_Prot/Parser.ipynb notebook.

How to do it...

Take a look at the following steps:

1. First, let's retrieve a file to work with. We will only retrieve, not parse as follows:

```
from __future__ import print_function
from Bio import PDB
repository = PDB.PDBList()
repository.retrieve_pdb_file('1TUP', pdir='.')
```

2. We will now devise a basic parsing framework that is capable of dealing with three
    types of records (on a single line, spanning multiple lines, and multiple records with
    the same name):

```
rec_types = {
    #single line
    'HEADER': [(str, 11, 49), (str, 50, 58), (str, 62,
65)],
    #multi_line
    'SOURCE': [(int, 7, 9), (str, 10, 78)],
    #multi_rec
    'LINK' : [(str, 12, 15), (str, 16, 16), (str, 17, 19),
(str, 21, 21), (int, 22, 25),
            (str, 26, 26), (str, 42, 45), (str, 46, 46),
(str, 47, 49), (str, 51, 51),
            (int, 52, 55), (str, 56, 56), (str, 59, 64),
(str, 66, 71), (float, 73, 77)],
    'HELIX': [(int, 7, 9), (str, 11, 13), (str, 15, 17),
(str, 19, 19), (int, 21, 24),
            (str, 25, 25), (str, 27, 29), (str, 31, 31),
            (int, 33, 36), (str, 37 ,37), (int, 38, 39),
(str, 40, 69), (int, 71, 75)],
    'SHEET': [(int, 7, 9), (str, 11, 13), (int, 14, 15),
(str, 17, 19), (str, 21, 21),
            (int, 22, 24), (str, 26, 26), (str, 28, 30),
            (str, 32, 32), (int, 33, 36), (str, 37, 37),
(int, 38, 39), (str, 41, 44),
            (str, 45, 47), (str, 49, 49), (int, 50, 53),
(str, 54, 54), (str, 56, 59),
            (str, 60, 62), (str, 64, 64), (int, 65, 68),
(str, 69, 69)],
}


def parse_pdb(hdl):
    for line in hdl:
        line = line[:-1]  # remove \n
        toks = []
        for section, elements in rec_types.items():
            if line.startswith(section):
                for fun, start, end in elements:
                    try:
                        toks.append(fun(line[start: end +
1]))
                    except ValueError:
```

```
                              toks.append(None)
                        yield (section, toks)
                  if len(toks) == 0:
                        yield ('UNKNOWN', line)
```

- ❑ Without contest, this is the ugliest piece of code in this book. The PDB format uses fixed size fields (for a reference of all records, refer to the previous recipe), so when we want to parse a certain record type, we need to hard code the positions of various fields. Here, we also type fields between string, integer, and float. You can extend this list using records that you need to extract.

- ❑ Finally, we have a function that will traverse the whole file as well, reporting unknown records with the complete line.

3. Let's parse our file as the first pass (this PDB file is suffixed as `.ent` because of Biopython's download procedure; PDB files normally end in `.pdb`):

```
hdl = open('pdb1tup.ent')
done_rec = set()
for rec in parse_pdb(hdl):
      if rec[0] == 'UNKNOWN' or rec[0] in done_rec:
            continue
      print(rec)
      done_rec.add(rec[0])
```

- ❑ We print the first instance of each record type and the output is shown in the following screenshot:

```
('HEADER', ['NTITUMOR PROTEIN/DNA              ', '11-JUL-95', '1TUP'])
('SOURCE', [None, 'MOL_ID: 1;                                     '])
('HELIX', [1, '  1', 'SER', 'A', 166, ' ', 'HIS', 'A', 168, ' ', 5, '                      ', 3])
('SHEET', [1, ' A', 4, 'ARG', 'A', 11, ' ', 'GLY', 'A', 112, ' ', 0, '   ', '  ', ' ', None, ' ', '   ', '  ', ' ', None, ' ']
)
('LINK', ['ZN  ', ' ', ' ZN', 'A', 951, ' ', ' SG ', ' ', 'CYS', 'A', 176, ' ', ' 1555', ' 1555', 2.33])
```

4. Let's now join records that span multiple lines using the following code:

```
multi_lines = ['SOURCE']

#assume multi is just a string
def process_multi_lines(hdl):
      current_multi = ''
      current_multi_name = None
      for rec_type, toks in parse_pdb(hdl):
            if current_multi_name is not None and current_multi_name \
                  != rec_type:
                  yield current_multi_name, [current_multi]
                  current_multi = ''
                  current_multi_name = None
```

```
            if rec_type in multi_lines:
                current_multi += toks[1].strip().rstrip() + ' '
                current_multi_name = rec_type
            else:
                if len(current_multi) != 0:
                    yield current_multi_name, [current_multi]
                    current_multi = ''
                    current_multi_name = None
                yield rec_type, toks
    if len(current_multi) != 0:
        yield current_multi_name, [current_multi]
```

- ❑ Here, we declare the SOURCE record as multiline. Our code will get all SOURCE lines and join them in a single string (for now, we assume that it's all a simple unstructured string).

- ❑ Of course, you can add more record types.

5. Again, we process using our example PDB file:

```
hdl = open('pdb1tup.ent')
done_rec = set()
for rec in process_multi_lines(hdl):
    if rec[0] == 'UNKNOWN' or rec[0] in done_rec:
        continue
    print(rec)
    done_rec.add(rec[0])
```

- ❑ The source is now a single entry, as shown in the following screenshot:

```
('HEADER', ['NTITUMOR PROTEIN/DNA                      ', '11-JUL-95', '1TUP'])
('SOURCE', ['MOL_ID: 1; SYNTHETIC: YES; MOL_ID: 2; SYNTHETIC: YES; MOL_ID: 3; ORGANISM_SCIENTIFIC: HOMO SAPIENS; ORGANISM_COMMON: H
UMAN; ORGANISM_TAXID: 9606; CELL_LINE: A431; CELL: HUMAN VULVA CARCINOMA; EXPRESSION_SYSTEM: ESCHERICHIA COLI; EXPRESSION_SYSTEM_TA
XID: 562; EXPRESSION_SYSTEM_PLASMID: PET3D '])
('HELIX', [1, '  1', 'SER', 'A', 166, ' ', 'HIS', 'A', 168, ' ', 5, '                  ', 3])
('SHEET', [1, '  A', 4, 'ARG', 'A', 11, ' ', 'GLY', 'A', 112, ' ', 0, '     ', '  ', '  ', ' ', None, ' ', '   ', '  ', ' ', ' ', None, ' ']
)
```

6. Finally, let's deal with more structured types. In this case, the content of SOURCE is known to be a specification list, as shown in the following code:

```
def get_spec_list(my_str):
    #ignoring escape characters
    spec_list = {}
    elems = my_str.strip().strip().split(';')
    for elem in elems:
        toks = elem.split(':')
        spec_list[toks[0].strip()] = toks[1].strip()
```

```
            return spec_list

    struct_types = {
        'SOURCE': [get_spec_list]
    }

    def process_struct_types(hdl):
        for rec_type, toks in process_multi_lines(hdl):
            if rec_type in struct_types.keys():
                funs = struct_types[rec_type]
                struct_toks = []
                for tok, fun in zip(toks, funs):
                    struct_toks.append(fun(tok))
                yield rec_type, struct_toks
            else:
                yield rec_type, toks
```

- ❑ This will parse a specification list. Refer to the preceding SOURCE output to get an idea of the content or the PDB specification:

7. Use the preceding code on our example PDB:

```
PDB:hdl = open('pdb1tup.ent'):
for rec in process_struct_types(hdl):
    if rec[0] != 'SOURCE':
        continue
    print(rec)
```

- ❑ The SOURCE is converted to an easy-to-process dictionary:

```
('SOURCE', [{'SYNTHETIC': 'YES', 'MOL_ID': '3', 'EXPRESSION_SYSTEM': 'ESCHERICHIA COLI', 'EXPRESSION_SYSTEM_TAXID': '562', 'ORGANIS
M_SCIENTIFIC': 'HOMO SAPIENS', 'CELL': 'HUMAN VULVA CARCINOMA', 'EXPRESSION_SYSTEM_PLASMID': 'PET3D', 'CELL_LINE': 'A431', 'ORGANIS
M_TAXID': '9606', 'ORGANISM_COMMON': 'HUMAN'}])
```

There's more...

This framework will help you parse a PDB file on Python should you need it. You should not be afraid of the format as it's of fixed length and simple to process. A poor man's parser can be something as simple as grep, in many cases, may work.

Animating with PyMol

Here, we will create a video of the p53 1TUP model. We will start our animation by going around the p53 1TUP model and then zooming in; as we zoom in, we change the render strategy so that you can see what is deep in the model. You can find the YouTube version of the video that you will generate at https://www.youtube.com/watch?v=CuEP40Id9O8.

Getting ready

This is the first case in which there is no IPython Notebook. Integration into IPython Notebook is not a problem, but as we will use Anaconda as a backend and PyMol does not play very well out of the box, we will use standard Python here. Assuming that you are using the Anaconda Python, if something does not work, I suggest you revert to the standard Python.

You will need to install PyMol (http://www.pymol.org); note that there is a free version as well. On Debian/Ubuntu/Linux, you can apt-get install pymol.

PyMol is more an interactive program than a Python library, so I strongly encourage you to play with it before moving on to the recipe. This can be fun!

The code for this recipe is available on the GitHub repository as a script (not as a notebook) along with chapter notebooks at notebooks/06_Prot. We will use the PyMol_Movie.py file.

How to do it...

Take a look at the following steps:

1. Let's initialize and retrieve our PDB model and prepare the rendering as follows:

```python
import pymol
from pymol import cmd
#pymol.pymol_argv = ['pymol', '-qc'] #  Quiet / no GUI
pymol.finish_launching()

cmd.fetch('1TUP', async=False)

cmd.disable('all')
cmd.enable('1TUP')
cmd.hide('all')
cmd.show('sphere', 'name zn')
```

   □ Note that the pymol_argv line makes the code silent. In your first executions, you may want to comment this out and see the user interface. For movie rendering, this will come in handy (as we will see soon).

   □ As a library, PyMol is quite tricky to use. For instance, after the import, you have to call finish_launching.

   □ We then fetch our PDB file.

   □ What then follows is a set of PyMol commands. Many web guides for interactive usage can be quite useful to understand what is going on. Here, we will enable all model for viewing purposes, hiding all (because the default view is lines and this is not good enough), then making zincs visible as spheres.

   □ At this stage, barring zincs, everything else is invisible.

2. In order to render our model, we will use three scenes as follows:

```
cmd.show('surface', 'chain A+B+C')
cmd.show('cartoon', 'chain E+F')
cmd.scene('S0', action='store', view=0, frame=0, animate=-1)

cmd.show('cartoon')
cmd.hide('surface')

cmd.scene('S1', action='store', view=0, frame=0, animate=-1)

cmd.hide('cartoon', 'chain A+B+C')
cmd.show('mesh', 'chain A')
cmd.show('sticks', 'chain A+B+C')
cmd.scene('S2', action='store', view=0, frame=0, animate=-1)
```

❑ We need to define two scenes. One scene is for the time when we go around the protein (surface-based, thus opaque) and the other is for the time when we dive in (cartoon-based). The DNA is always a cartoon.

❑ We also define a third scene for when we zoom out at the end. The protein will get rendered as sticks, and we add a mesh to chain A so that the relationship becomes clearer with the DNA.

3. Let's define the basic parameter of our video as follows:

```
cmd.set('ray_trace_frames', 0)
cmd.mset(1, 500)
```

❑ We define the default ray tracing algorithm. This line does not need to be there, but try to increase the number to 1, 2, or 3 and be ready to wait a lot.

❑ You can only use 0 if you have the OpenGL interface on (with the GUI), so, for this fast version, you will need to have the GUI on (the preceding pymol_ argv should be commented as it is).

❑ We then inform PyMol that we will have 500 frames.

4. In the first 150 frames, we move around using the initial scene. We go around the model a bit, and go near the DNA using the following code:

```
cmd.frame(0)
cmd.scene('S0')
cmd.mview()
cmd.frame(60)
cmd.set_view((-0.175534308,    -0.331560850,    -0.926960170,
              0.541812420,      0.753615797,    -0.372158051,
              0.821965039,     -0.567564785,     0.047358301,
              0.000000000,      0.000000000,  -249.619018555,
             58.625568390,     15.602619171,    77.781631470,
            196.801528931,    302.436492920,   -20.000000000))


cmd.mview()
cmd.frame(90)
cmd.set_view((-0.175534308,    -0.331560850,    -0.926960170,
              0.541812420,      0.753615797,    -0.372158051,
              0.821965039,     -0.567564785,     0.047358301,
             -0.000067875,      0.000017881,  -249.615447998,
             54.029174805,     26.956727982,    77.124832153,
            196.801528931,    302.436492920,   -20.000000000))
cmd.mview()
cmd.frame(150)
cmd.set_view((-0.175534308,    -0.331560850,    -0.926960170,
              0.541812420,      0.753615797,    -0.372158051,
              0.821965039,     -0.567564785,     0.047358301,
             -0.000067875,      0.000017881,   -55.406421661,
             54.029174805,     26.956727982,    77.124832153,
              2.592475891,    108.227416992,   -20.000000000))
cmd.mview()
```

- ❏ We define three points; the first two align with the DNA and the last goes in.
- ❏ The way to get coordinates (all these numbers) is to use PyMol in the interactive mode, navigate using the mouse and keyboard, and use the `get_view` command, which will return coordinates that you can cut and paste.

❏   The first frame is as follows:

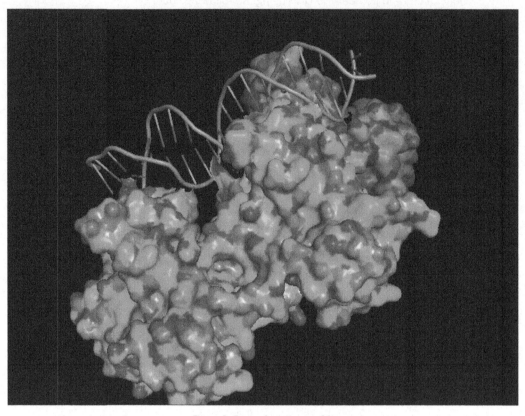

Figure 2: Frame 0 and scene S0

5. We now change the scene, in preparation to go inside the protein:

```
cmd.frame(200)
cmd.scene('S1')
cmd.mview()
```

❑　The following figure shows the current position:

Figure 3: Frame 200 near the DNA molecule and scene S1

6.　We move inside the protein and change the scene at the end using the following code:

```
cmd.frame(350)
cmd.scene('S1')
cmd.set_view ((0.395763457,    -0.173441306,     0.901825786,
               0.915456235,     0.152441502,    -0.372427106,
              -0.072881661,     0.972972929,     0.219108686,
               0.000070953,     0.000013039,   -37.689743042,
              57.748500824,    14.325904846,    77.241867065,
             -15.123448372,    90.511535645,   -20.000000000))

cmd.mview()
cmd.frame(351)
cmd.scene('S2')
cmd.mview()
```

❑ We are now full in, as shown in the following figure:

Figure 4: Frame 350, scene S1 on the verge of changing to S2

7. Finally, we let PyMol return to its original position, play, save, and quit:

```
cmd.frame(500)
cmd.scene('S2')
cmd.mview()
cmd.mplay()
cmd.mpng('p53_1tup')

cmd.quit()
```

❑ This will generate 500 PNG files with the p53_1tup prefix.

❑   Here is a frame approaching the end (450):

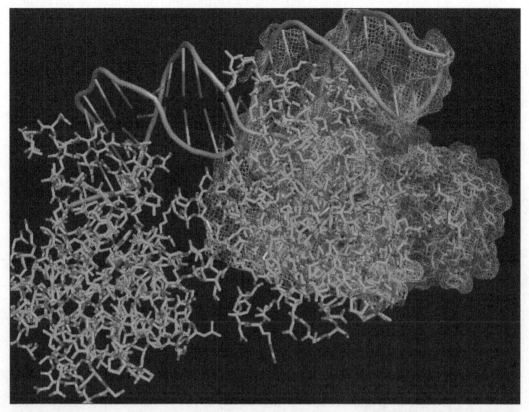

Figure 5: Frame 450 and scene S2

There's more...

The YouTube video was generated using Libav's `avconv` on Linux at 15 frames per second as follows:

```
avconv -r 15  -f image2 -start_number 1  -i "p53_1tup%04d.png"  \
example.mp4
```

There are plenty of applications used to generate videos from images. PyMol can generate an MPEG, but this means we have to install extra libraries.

PyMol was created to be used interactively from its console (which can be extended in Python). Using it the other way around (importing from Python with no GUI) can be complicated and frustrating; PyMol starts a separate thread to render images that works asynchronously. For example, this means that your code may be in a different position from where the renderer is. I have put another script called `PyMol_Intro.py` on the GitHub repository; you will see that the second PNG call will start before the first one has finished. Try the following code and see how you expect it to behave, and how it actually behaves.

There is plenty of good documentation for PyMol from a GUI perspective at `http://www.pymolwiki.org/index.php/MovieSchool`. This is a great starting point if you want to make movies and `http://www.pymolwiki.org` is a treasure trove of information.

Parsing mmCIF files using Biopython

The mmCIF file format is probably the future. Biopython does not have yet full functionality to work with it, but we will take a look at what is here now.

Getting ready

As Bio.PDB is not able to automatically download mmCIF files, you need to get your protein file and rename it as 1tup.cif. This can be found at `https://github.com/tiagoantao/bioinf-python/blob/master/notebooks/Datasets.ipynb` under the `1TUP.cif` name.

You can find this content in the `06_Prot/mmCIF.ipynb` notebook.

How to do it...

Take a look at the following steps:

1. Let's parse the file. We just use the mmCIF parser instead of the PDB parser:

```
from __future__ import print_function
from Bio import PDB
parser = PDB.MMCIFParser()
p53_1tup = parser.get_structure('P53', '1tup.cif')
```

2. Let's inspect the following chains:

```
def describe_model(name, pdb):
    print()
    for model in p53_1tup:
        for chain in model:
            print('%s - Chain: %s. Number of residues: %d. Number of atoms: %d.' %
                                    (name, chain.id, len(chain),
                                     len(list(chain.get_atoms()))))
describe_model('1TUP', p53_1tup)
```

- ❑ Note that we have new chains for exactly the same model, as shown in the following figure:

```
1TUP - Chain: A. Number of residues: 21. Number of atoms: 420.
1TUP - Chain: B. Number of residues: 21. Number of atoms: 435.
1TUP - Chain: C. Number of residues: 196. Number of atoms: 1535.
1TUP - Chain: D. Number of residues: 194. Number of atoms: 1522.
1TUP - Chain: E. Number of residues: 195. Number of atoms: 1529.
1TUP - Chain: F. Number of residues: 1. Number of atoms: 1.
1TUP - Chain: G. Number of residues: 1. Number of atoms: 1.
1TUP - Chain: H. Number of residues: 1. Number of atoms: 1.
1TUP - Chain: I. Number of residues: 198. Number of atoms: 198.
1TUP - Chain: J. Number of residues: 70. Number of atoms: 70.
1TUP - Chain: K. Number of residues: 80. Number of atoms: 80.
1TUP - Chain: L. Number of residues: 22. Number of atoms: 22.
1TUP - Chain: M. Number of residues: 14. Number of atoms: 14.
```

3. Let's check what's inside these new chains:

```
done_chain = set()
for residue in p53_1tup.get_residues():
    chain = residue.parent
    if chain.id in done_chain:
        continue
    done_chain.add(chain.id)
    print(chain.id, residue.id)
```

- ❑ So, zincs and waters were unpacked for proteins and DNA
- ❑ Note the water ID; if you have code working to detect water by getting the W residue in the PDB file, you may have to perform a slight change here

4. Many of the fields are not available on the parsed structure, but it can still be retrieved using a lower-level dictionary as follows:

```
mmcif_dict = PDB.MMCIF2Dict.MMCIF2Dict('1tup.cif')
for k, v in mmcif_dict.items():
    print(k, v)
    print()
```

- ❑ Unfortunately, this list is large and requires some postprocessing to make some sense of it, but this is available.

There's more...

You still have all the model information from the mmCIF file made available by Biopython, so the parser is still quite useful. You can expect more developments with the mmCIF parser than with the PDB parser.

There is a Python library made available by the PDB at `http://mmcif.wwpdb.org/docs/sw-examples/python/html/index.html`.

8

Other Topics in Bioinformatics

In this chapter, we will cover the following recipes:

- ▸ Accessing the Global Biodiversity Information Facility via REST
- ▸ Georeferencing GBIF datasets
- ▸ Accessing molecular interaction databases with PSIQUIC
- ▸ Plotting protein interactions with Cytoscape the hard way

Introduction

In this chapter, we will address some topics that are well within the remit of computational biology and deserve at least some reference. We will start with two recipes using the **Global Biodiversity Information Facility** (**GBIF**), a database of worldwide scientific data on biodiversity. After this, we will interface Python with Cytoscape, a powerful software platform to visualize genomic and proteomic interaction networks. To perform visualization with Cytoscape, we will first set the stage by accessing the PSIQUIC service, a common query interface to several molecular interaction databases.

We will take the opportunity to indirectly introduce other topics, such as more bioinformatics databases, graph processing, and geo-referencing that are relevant in our context (one way or another). To interface, we will use only REST APIs to all databases and services, making the code in all recipes here quite streamlined. You may want to refresh your knowledge of REST architectures. We will use the requests library for REST interfacing. If you never used it, do not worry, it's actually quite easy. We will also use the IPython Notebook facilities.

Accessing the Global Biodiversity Information Facility

The **Global Biodiversity Information Facility** (**GBIF**), http://www.gbif.org, makes the available information about biodiversity in a programmatic friendly way using a REST API. In GBIF, we will find evidence for occurrence of species across the planet and much of this information is geo-referenced.

In this recipe, we will concentrate on two types of GBIF information: species and occurrences. Species are actually a more general taxonomic framework and occurrences record observations of species.

In this recipe, we will try to extract the biodiversity information related to bears. You can find this content in the 07_Other/GBIF.ipynb notebook.

How to do it...

Take a look at the following steps:

1. First, let's define a function to get the data on REST, as shown in the following code:

```
from __future__ import print_function
import requests
def do_request(service, a1=None, a2=None, a3=None,
    **kwargs):
    server = 'http://api.gbif.org/v1'
    params = ''
    for a in [a1, a2, a3]:
        if a is not None:
            params += '/' + a
    req = requests.get('%s/%s%s' % (server, service,
        params),
                            params=kwargs,
                            headers={'Content-Type':
                                            'application/json'})
    if not req.ok:
        req.raise_for_status()
    return req.json()
```

2. Then, see how many species records refer to the 'bear' word. Remember that this is actually more general than species. You will also get records for all kinds of taxonomic ranks with the following code:

```
req = do_request('species', 'search', q='bear')
print(req['count'])
```

As of today, we have 19,204 records. A typical record contains information about taxonomic rank, the relevant taxonomic information (kingdom, family, genus, and so on), and the scientific rank. The following screenshot is an example of part of a record:

```
{u'authorship': u'',
 u'canonicalName': u'Ursus white bear',
 u'class': u'Mammalia',
 u'classKey': 106223020,
 u'datasetKey': u'fab88965-e69d-4491-a04d-e3198b626e52',
 u'descriptions': [],
 u'family': u'Ursidae',
 u'familyKey': 106657396,
 u'genus': u'Ursus',
 u'genusKey': 106658119,
 u'habitats': [],
 u'higherClassificationMap': {u'106148414': u'Metazoa',
  u'106151875': u'Carnivora',
  u'106223020': u'Mammalia',
  u'106522535': u'Chordata',
  u'106657396': u'Ursidae',
  u'106658119': u'Ursus'},
 u'key': 106189791,
 u'kingdom': u'Metazoa',
 u'kingdomKey': 106148414,
 u'nameType': u'SCINAME',
 u'nomenclaturalStatus': [],
 u'numDescendants': 0,
 u'numOccurrences': 0,
 u'order': u'Carnivora',
 u'orderKey': 106151875,
 u'parent': u'Ursus',
 u'parentKey': 106658119,
 u'phylum': u'Chordata',
 u'phylumKey': 106522535,
 u'rank': u'SPECIES',
 u'scientificName': u'Ursus sp. Shennongjia white bear',
 u'species': u'Ursus white bear',
```

3. Almost 20,000 records is a bit too much to inspect; let's restrict ourselves to the ones that are on the taxonomic rank of a family, as shown in the following code:

```
req_short = do_request('species', 'search', q='bear',
    rank='family')
print(req_short['count'])
bear = req_short['results'][0]
```

These 645 records are ordered by relevance to the search term; according to GBIF's algorithms, we will take the very first record to continue our work.

4. As GBIF limits the number of records that you can get per call, let's use the following function to get all records (of course, to be used with care):

```python
import time
def get_all_records(rec_field, service, a1=None, a2=None,
        a3=None, **kwargs):
    records = []
    all_done = False
    offset = 0
    num_iter = 0
    while not all_done and num_iter < 100:  # arbitrary
        req = do_request(service, a1=a1, a2=a2, a3=a3,
            offset=offset, **kwargs)
        all_done = req['endOfRecords']
        if not all_done:
            time.sleep(0.1)
        offset += req['limit']
        records.extend(req[rec_field])
        num_iter += 1
    return records
```

❑ Many REST services offer paging APIs, that is, you can take results in parts by issuing multiple calls, specifying the starting point (in this case, `offset`) and a limit of the number of results. As we want to be good citizens, we include a sleep of 0.1 seconds between calls to the server so that there is no excessive burden on it.

5. Now, given a certain internal node in the taxonomic tree, let's get all the leaves down from that node, as shown in the following code:

```python
def get_leaves(nub):
    leaves = []
    recs = get_all_records('results', 'species', str(nub),
        'children')
    if len(recs) == 0:
        return None
    for rec in recs:
        rec_leaves = get_leaves(rec['nubKey'])
        if rec_leaves is None:
            leaves.append(rec)
        else:
            leaves.extend(rec_leaves)
    return leaves
```

❑ Here, `nub` is GBIF's taxonomy identifier.

6. Finally, let's get the leaves for the first bear node that we got on the search by the `'family'` rank. We will mostly have the name and taxonomy information. We print the scientific name, rank of the record, and the vernacular name if it exists for all the leaves:

```
records = get_all_records('results', 'species',
    str(bear['nubKey']), 'children')
leaves = get_leaves(bear['nubKey'])
for rec in leaves:
    print(rec['scientificName'], rec['rank'], end=' ')
    vernaculars = do_request('species', str(rec['nubKey']),
        'vernacularNames', language='en')['results']
    for vernacular in vernaculars:
        if vernacular['language'] == 'eng':
            print(vernacular['vernacularName'], end='')
            break
    print()
```

❑ Note that the vernacular name comes from another service. GBIF has vernacular names in many languages; here, we will choose the English version. This call will take a few seconds because of our sleep code introduced before.

7. For all leaves, we will now summarize the source of all records, the country of observation, the number of extinct references, and the species with no occurrences at all (remember that occurrence is another fundamental GBIF concept), as shown in the following code:

```
from collections import defaultdict
basis_of_record = defaultdict(int)
country = defaultdict(int)
zero_occurrences = 0
count_extinct = 0
for rec in leaves:
    occurrences = get_all_records('results', 'occurrence',
        'search', taxonKey=rec['nubKey'])
    for occurrence in occurrences:
        basis_of_record[occurrence['basisOfRecord']] += 1
        country[occurrence.get('country', 'NA')] += 1
    if len(occurrences) > 0:
        zero_occurrences += 1
    profiles = do_request('species', str(rec['nubKey']),
        'speciesProfiles')['results']
    for profile in profiles:
        if profile.get('extinct', False):
            count_extinct += 1
            break
```

❏ We maintain two dictionaries. One is `basis_of_record` with a count per different record origin. Second is `country` with a count per country of observation (there is also the country that published the results and it's different many times) We also check the `speciesProfiles` service to see whether a record is labeled as `extinct`.

8. Let's plot this as follows:

```
import numpy as np
import matplotlib.pyplot as plt

countries, obs_countries = zip(*sorted(country.items(),
    key=lambda x: x[1]))
basis_name, basis_cnt = zip(*sorted(basis_of_record.items(),
    key=lambda x: x[1]))
fig = plt.figure(figsize=(16, 9))
ax = fig.add_subplot(1, 2, 1)
ax.barh(np.arange(10) - 0.5, obs_countries[-10:])
ax.set_title('Top 10 countries per occurrences')
ax.set_yticks(range(10))
ax.set_ylim(0.5, 9.5)
ax.set_yticklabels(countries[-10:])

ax = fig.add_subplot(2, 2, 2)
ax.set_title('Basis of record')
ax.bar(np.arange(len(basis_name)), basis_cnt, color='g')
basis_name = [x.replace('OBSERVATION',
    'OBS').replace('_SPECIMEN', '') for x in basis_name]
ax.set_xticks(0.5 + np.arange(len(basis_name)))
ax.set_xticklabels(basis_name, size='x-small')

ax = fig.add_subplot(2, 2, 4)
other = len(leaves) - zero_occurrences - count_extinct
pie_values = [zero_occurrences, count_extinct, other]
labels = ['No occurrence (%d)' % zero_occurrences,
        'Extinct (%d)' % count_extinct, 'Other (%d)' % other]
ax.pie(pie_values, labels=labels,
    colors=['cyan', 'magenta', 'yellow'])
ax.set_title('Status for each species')
```

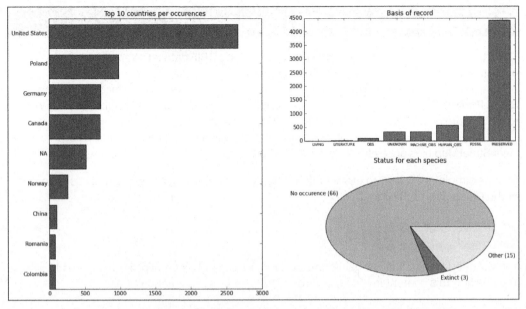

Figure 1: Most referred countries of occurrence, the distribution of the origin of records,
and the status of the record in terms of extinction and occurrence

❑ Be careful with pie charts because they are seen by many as difficult to
interpret. Here, we explicitly added the number of observations per group
to make ours clearer.

❑ If you prefer, you can import seaborn with matplotlib to get a more modern
look. There are quite a few examples of seaborn throughout the book.

❑ Note that some records do not have country information (the NA entry on
the chart).

There's more...

The GBIF database seems not to be totally consistent in terms of data. The sources
of the occurrences and species differ. Therefore, there is not too much effort put in to
standardization. From species names to information about countries (such as the preceding
NA case), you can see this in different parts of the database. Also, many records do not have
the complete geo-referenced data, so be careful when analyzing the data. The REST API is
documented at `http://www.gbif.org/developer/summary`.

Geo-referencing GBIF datasets

Here, we will work with the geo-referenced data from the GBIF dataset. We will take this opportunity to see how to interface with OpenStreetMap (https://www.openstreetmap.org), a freely available mapping service. We will also use a Python image processing library called Pillow (http://python-pillow.github.io/, which is based on PIL). You may want to read a little bit on both before starting. Tile Map Services (http://wiki.openstreetmap.org/wiki/TMS) will be quite an important concept to get a basic grasp of. GBIF and OpenStreetMap tiles are available behind REST services.

In our example, we will try to extract information from GBIF using the geographic coordinates of the Galápagos archipelago.

Getting ready

You will need to install Pillow using `conda install pillow` or `pip install pillow`. You can find this content in the `07_Other/GBIF_extra.ipynb` notebook.

How to do it...

Take a look at the following steps:

1. First, let's define a function to get a map tile (a PNG image of size 256 x 256) from OpenStreetMap. We will also define a function to convert geographical coordinate systems to tile indexes, as shown in the following code:

```python
from __future__ import division, print_function
import math
import requests
def get_osm_tile(x, y, z):
    url = 'http://tile.openstreetmap.org/%d/%d/%d.png' % (z, x, y)
    req = requests.get(url)
    if not req.ok:
        req.raise_for_status()
    return req

def deg_xy(lat, lon, zoom):
    lat_rad = math.radians(lat)
    n = 2 ** zoom
    x = int((lon + 180) / 360 * n)
    y = int((1 - math.log(math.tan(lat_rad) + (1 / math.cos(lat_rad))) / math.pi) / 2 * n)
    return x, y
```

- ❏ This code is responsible for getting a tile from the OpenStreetMap server, which is a REST service. The most complicated part to understand in this recipe is the conversion between geographical coordinates and the tiling system, so `deg_xy`, which converts latitude, longitude, and zoom to the tile coordinates. Let's use this to make this clear.

2. Let's get four tiles at different zoom levels around our coordinates of interest. We will then compose a single image with Pillow (including all tiles) as follows:

```
import sys
import PIL.Image
if sys.version_info.major == 2:
    from StringIO import StringIO
else:
    from io import StringIO
from IPython import display


lat, lon = -0.666667, -90.55


pils = []
for zoom in [0, 1, 5, 8]:
    x, y = deg_xy(lat, lon, zoom)
    print(x,y,zoom)
    osm_tile = get_osm_tile(x, y, zoom)
    pil_img = PIL.Image.open(StringIO(osm_tile.content))
    pils.append(pil_img)
composite = PIL.Image.new('RGBA', (520, 520))
print(pils[0].mode, pils[0].size)
composite.paste(pils[0], (0, 0, 256, 256))
composite.paste(pils[1], (264, 0, 520, 256))
composite.paste(pils[2], (0, 264, 256, 520))
composite.paste(pils[3], (264, 264, 520, 520))
```

- ❏ We will use the latitude and longitude extracted from Wikipedia for the Galápagos.

- ❏ We will work at four zoom levels; here, 0 means the whole world. In this case, there is only a single tile with x = 0 and y = 0 coordinates. We then take the zoom level of 1 with has a total of 4 tiles (2 x 2) for the planet and we use the tile with coordinates x = 0 and y = 1. We then go to level 5, where we have 1024 tiles (32 x 32 or 2**5) with a x = 7 and y = 16. Finally, we take the level 8 of zoom with 65536 tiles (256 x 256) with x = 63 and y = 128.

- ❏ We take the four tiles, one for each zoom level, which are of 256 x 256 resolution and then use Pillow to compose an image.

3. We now define a function to convert Pillow images to IPython Notebook images and display it:

```
from io import BytesIO
def convert_pil(img):
    b = BytesIO()
    img.save(b, format='png')
    return display.Image(data=b.getvalue())
convert_pil(composite)
```

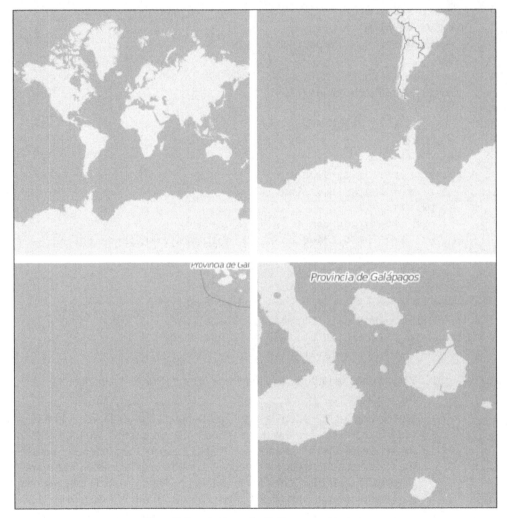

Figure 2: Zooming in on the Galápagos archipelago

❏ Note that at a zoom level of five and eight, it's impossible to get the whole archipelago within a single tile. Let's solve this in the next step.

4. Let's define a function to get the surrounding tiles. This function will abstract away the source of all tiles. With this, we can use the same function with OpenStreetMap and other servers (such as the GBIF server). Again, we will use Pillow to join several tiles, as shown in the following code:

```
def get_surrounding(x, y, z, tile_fun):
    composite = PIL.Image.new('RGBA', (768, 768))
    for xi, x_ in enumerate([x - 1, x, x + 1]):
        for yi, y_ in enumerate([y - 1, y, y + 1]):
            tile_req = tile_fun(x_, y_, z)
            pos = (xi * 256, yi * 256, xi * 256 + 256, yi *
                256 + 256)
            img = \
PIL.Image.open(StringIO(tile_req.content))
            composite.paste(img, pos)
    return composite
```

5. Let's get the tiles for the area around the Galápagos. Therefore, we get 3 x 3 tiles in a single 768 x 768 image:

```
zoom = 8
x, y = deg_xy(lat, lon, zoom)
osm_big = get_surrounding(x, y, zoom, get_osm_tile)
```

6. Finally, let's get the GBIF tile. We start with getting the worldwide tile (zoom of 0) for our bears from the preceding recipe. We will not plot these here, but you can easily see the result from the preceding code in the corresponding notebook:

```
def get_gbif_tile(x, y, z, **kwargs):
    server = 'http://api.gbif.org/v1'
    kwargs['x'] = str(x)
    kwargs['y'] = str(y)
    kwargs['z'] = str(z)
    req = requests.get('%s/map/density/tile' % server,
                       params=kwargs,
                       headers={})
    if not req.ok:
        req.raise_for_status()
    return req
gbif_tile = get_gbif_tile(0, 0, 0, resolution='4',
                    type='TAXON', key='6163845')


img = PIL.Image.open(StringIO(gbif_tile.content))
```

- ❑ This code is very similar to the OpenStreetMap one because they are both REST-based.

- ❑ The `img` variable contains a Pillow representation of the 256 x 256 zoom 0 level for GBIF on our bears. The GBIF interface allows you to constrain the result in many ways; here, we want our bear tax on ID.

7. It turns out that it's quite easy to query GBIF for all the occurrences and species in our Galápagos tiles, as shown in the following code:

```
import functools
zoom = 8
x, y = deg_xy(lat, lon, zoom)
gbif_big = get_surrounding(x, y, zoom,
                                    functools.partial(get_
gbif_tile,
                                    hue='0.1',
                                    resolution='2',
                                    saturation='True'))
```

- ❑ Note that the code here is remarkably similar to the previous application of `get_surrounding`. Indeed, the most complex piece of code has nothing to do with GBIF; we perform a partial function application to instantiate some GBIF parameters regarding visual parameters.

8. Now we will use Pillow again to join images from OpenStreetMap and GBIF:

```
compose = PIL.Image.alpha_composite(osm_big, gbif_big)
convert_pil(compose)
```

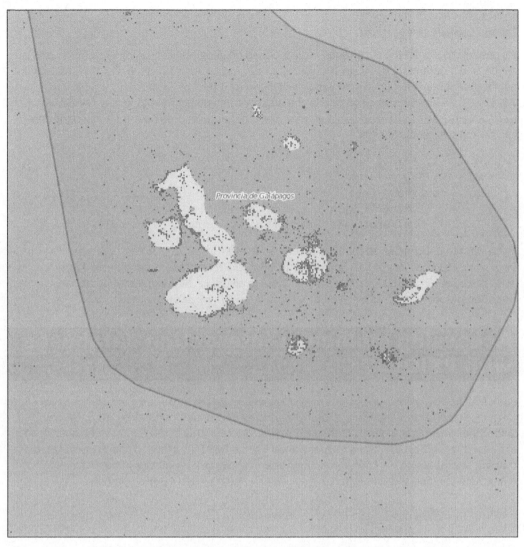

Figure 3: Overlaying species information from GBIF on top of OpenStreetMap tiles

There's more...

It's possible to extract the occurrence records based on geographical coordinates. In the previous recipe, add a parameter `geometry` to your `do_request/get_all_records` call. This should be a textual representation in a subset of well-known text (`http://en.wikipedia.org/wiki/Well-known_text`). For details of the supported subset, refer to `http://www.gbif.org/developer/occurrence`. As an example, a rectangular area near the Galápagos area can be represented as follows:

```
start = 2, -93
end = 1, -91
geom = 'POLYGON(({xi} {yi}, {xf} {yi}, {xf} {yf}, {xi} {yf}, {xi}
    {yi}))'.format(
            xi=start[1], xf=end[1], yi=start[0], yf=end[0])
```

If you want to know more about tiling coordinates, refer to `http://wiki.openstreetmap.org/wiki/Slippy_map_tilenames`. The GBIF REST interface is documented at `http://www.gbif.org/developer/summary`. There is also a lot of documentation about OpenStreetMap; refer to its wiki link at `https://wiki.openstreetmap.org`.

Accessing molecular-interaction databases with PSIQUIC

PSIQUIC (`http://www.ebi.ac.uk/Tools/webservices/psicquic/view/main.xhtml`) is a consistent interface to many molecular-interaction databases. It's used inside Cytoscape, which is the object of the next recipe. We take this as an opportunity to learn how to interact with PSIQUIC. In this recipe, we will use its REST interface to check the databases that are available and perform some basic querying. We will revisit this in the Cytoscape recipe. You can find this content in the `07_Other/PSICQUIC.ipynb` notebook.

How to do it...

Take a look at the following steps:

1. First, let's define a convenient REST function as follows:

```
from __future__ import print_function
import requests
def get_psiquic(service, query, full_url=False, **kwargs):
    kwargs['format'] = kwargs.get('format', 'tab27')
    if full_url:
        req = requests.get('%s%s' % (service, query),
            params=kwargs)
    else:
```

```
        server = \
'http://www.ebi.ac.uk/Tools/webservices/psicquic'
        req = requests.get('%s/%s/%s' % (server, service,
            query), params=kwargs)
    if not req.ok:
        req.raise_for_status()
    return req.content
```

❏ This is a standard REST call, except for the fact that the default format is something called tab27, which is PSIQUIC-specific. We will revisit this in the next recipe.

2. We want to know which databases are registered and active, as shown in the following code:

```
import xml.etree.ElementTree as ET
import pandas as pd
def get_databases(db_xml):
    for service in db_xml:
        for elem in service:
            ns_clean_tag = elem.tag[elem.tag.find('}') +
                1:]
            if ns_clean_tag == 'name':
                name = elem.text
            elif ns_clean_tag == 'active':
                active = False if elem.text == 'false' else \
                        True
            elif ns_clean_tag == 'restUrl':
                rest_url = elem.text
            elif ns_clean_tag == 'restExample':
                example = elem.text
            elif ns_clean_tag == 'organizationUrl':
                org_url = elem.text
            else:
                pass  # there are a few more
        yield {'name': name, 'active': active, 'org_url':
                org_url,
                'example': example, 'rest_url': rest_url}

dbs_xml = get_psiquic('registry', 'registry',
    action='STATUS', format='xml')
dbs_xml_parsed = ET.fromstring(dbs_xml)
dbs = pd.DataFrame.from_records(get_databases(dbs_xml_parsed))

pd.options.display.max_colwidth = 100
active_dbs = dbs[dbs.active==True]
```

❑ We get the output of the registry service in the XML format, and parse the XML file using `xml.etree.ElementTree`. PSICQUIC makes available information (such as the service name, its URL, and if it's currently active or not). It also makes available examples of REST queries. We use pandas to get the active services.

3. Let's count all the records in all the databases referred to as *TP53* and then split the count by database, as shown in the following code:

```
req = \
get_psiquic('intact/webservices/current/search/query',
    'tp53', format='count')
print(req)
for index, db in active_dbs.iterrows():
    req = get_psiquic(db['rest_url'], 'query/tp53',
        full_url=True, format='count')
    count = int(req)
    print('DB: %s, count: %d' % ( db['name'], count))
```

❑ Part of the output is shown here. As with all online databases, this may change when you run it. The PSIQUIC case can be particularly variable, because it's a federation of databases:

```
4748
DB: BioGrid, count: 2375
DB: bhf-ucl, count: 14
DB: ChEMBL, count: 80
DB: DIP, count: 0
DB: HPIDb, count: 68
DB: InnateDB, count: 136
DB: IntAct, count: 4748
DB: mentha, count: 3218
DB: MPIDB, count: 0
DB: MatrixDB, count: 0
DB: MINT, count: 2158
DB: Reactome, count: 0
DB: Reactome-FIs, count: 369
DB: STRING, count: 2118
DB: BIND, count: 47
DB: Interoporc, count: 0
DB: I2D-IMEx, count: 194
DB: InnateDB-IMEx, count: 5
DB: MolCon, count: 18
DB: UniProt, count: 438
```

4. Finally, let's query 1000 records from all databases for *TP53* and see the types of records that we have. We will use the `csv` module here to deal with the return; you can also perform this with pandas:

```
import csv
req = get_psiquic('intact/webservices/current/search/query',
                                'tp53',
                                firstResult=0, maxResults=1000)
answer = csv.reader(StringIO(req), delimiter='\t')
db_types = set()
for record in answer:
    db_types.add(record[0].split(':')[0])
    db_types.add(record[1].split(':')[0])
print(db_types)
```

At most, we query 1000 records. PSIQUIC has a REST paging architecture (see the GBIF architecture for another example of paging). Also, we will query across all the databases. We then look at the record IDs in the result. For PSIQUIC, as it federates several databases, the first part of the identifier which specifies the database comes from the ID. For example, `uniprotkb:P04637` or `ensembl:ENST00000316024`.

In this result set, we have the `uniprotkb, ddbj/embl/genbank, -, ensembl, intact,` and `chebi` types.

We will inspect the content of a similar result in the next recipe.

Plotting protein interactions with Cytoscape the hard way

Cytoscape (`http://cytoscape.org/`) is a platform to visualize molecular interaction networks. Here, we will interact with Cytoscape using a REST interface. There are easier ways to perform this recipe, but we will take this opportunity to continue interacting with the PSICQUIC service. Also, we will exercise the `NetworkX` graph processing library (`https://networkx.github.io/`), which will be useful on its own.

Taking a page from *Chapter 7, Using the Protein Data Bank*, we will plot p53 interactions stored in the `UniProt` database.

Getting ready

You will need to install the Cytoscape version 3.2.1 (or higher), which will require Java 7 or preferably 8. You will also need the `cyREST` application in Cytoscape (see the **Apps** main menu in Cytoscape for this). The code will use a REST interface to communicate with Cytoscape, so it will run outside it, but it will require Cytoscape to be running, so start Cytoscape with cyREST before running the following code.

You should also install the py2cytoscape (via `pip`) and `NetworkX` (via `conda` or `pip`) Python libraries. You can find this content in the `07_Other/Cytoscape.ipynb` notebook. As this will require a massive download of software, the notebook will not work in our Docker implementation.

How to do it...

Take a look at the following steps:

1. First, let's access the PSICQUIC service (the `UniProt` database) via its REST interface, as shown in the following code:

```
from __future__ import print_function
import requests
def get_psiquic_uniprot(query, **kwargs):
    kwargs['format'] = kwargs.get('format', 'tab27')
    server = 'http://www.ebi.ac.uk/Tools/webservices/psicquic/
uniprot/webservices/current/search/query'
    req = requests.get('%s/%s' % (server, query),
        params=kwargs)
    return req.content
```

2. Then, get all the genes referred to (along with their respective species) and the interactions:

```
from collections import defaultdict
genes_species = defaultdict(set)
interactions = {}

def get_gene_name(my_id, alt_names):
    toks = alt_names.split('|')
    for tok in toks:
        if tok.endswith('(gene name)'):
            return tok[tok.find(':') + 1: tok.find('(')]
    return my_id + '?'  # no name...

def get_vernacular_tax(tax):
    return tax.split('|')[0][tax.find('(') + 1:-1]

def add_interactions(species):
    for rec in species.split('\n'):
        toks = rec.rstrip().split('\t')
        if len(toks) < 15:
            continue  # empty line at the end
        id1 = toks[0][toks[0].find(':') + 1:]
        id2 = toks[1][toks[1].find(':') + 1:]
```

```
        gene1, gene2 = get_gene_name(id1, toks[4]), \
                       get_gene_name(id2, toks[5])

        tax1, tax2 = get_vernacular_tax(toks[9]), get_vernacular_
tax(toks[10])
        inter_type = toks[11][toks[11].find('(') + 1:-1]
        miscore = float(toks[14].split(':')[1])
        genes_species[tax1].add(gene1)
        genes_species[tax2].add(gene2)
        interactions[((tax1, gene1), (tax2, gene2))] = \
            {'score': miscore, 'type': inter_type}
```

- ❑ We will create a dictionary with species as a key with a set of genes referred. We will also create a dictionary of interactions with the key being a tuple with the genes involved, the value, the iteration type, and miscore.

- ❑ The `add_interactions` changes a couple of external dictionaries in place. This dialect is not scalable in a very complex program because it is bug-prone. It works well for our small script, but be sure to adapt the code if you are using this in a larger infrastructure.

- ❑ The result is outputted in the PSI-MI TAB 2.7 format (`https://code.google.com/p/psimi/wiki/PsimiTab27Format`). This is an easy to parse and tab-delimited format, which includes the identifiers for both iterators (including the database), aliases (for example, gene names), the interaction method, taxonomy identifiers, and so on. Here, we take aliases, species, and the confidence score for each interaction.

3. Let's extract interactions with the human p53 protein, which has also homologous on rats and mice. You can discover the following IDs using `UniProt` or the code from the previous chapter:

```
human = get_psiquic_uniprot('uniprotkb:P04637')
add_interactions(human)
rat = get_psiquic_uniprot('uniprotkb:P10361')
add_interactions(rat)
mouse = get_psiquic_uniprot('uniprotkb:P02340')
add_interactions(mouse)
```

4. With this information, we can now start drawing on Cytoscape. Remember that Cytoscape must be running on the local machine and that the `cyREST` plugin must be installed. We will first construct a `NetworkX` graph. This graph will have extra annotations for genes and species added to the nodes and interaction types, and confidence scores added to the edges as follows:

```
import networkx as nx
server = 'http://localhost:1234/v1'
def get_node_id(species, gene):
    if species == 'human':
```

```
                    return gene
          elif species in ['mouse', 'rat']:
              return '%s (%s)' % (gene, species[0])
          else:
              return  '%s (%s)' % (gene, species)
    graph = nx.Graph()
    for species, genes in genes_species.items():
        for gene in genes:
            name = get_node_id(species, gene)
            graph.add_node(get_node_id(species, gene),
                            species=species, gene=gene)
    for (i1, i2), attribs in interactions.items():
        tax1, gene1 = i1
        tax2, gene2 = i2
        graph.add_edge(get_node_id(tax1, gene1),
                        get_node_id(tax2, gene2),
                        interaction=attribs['type'],
                        score=attribs['score'])
```

5. Now, let's now convert the `NetworkX` graph to a Cytoscape representation and plot it, as shown in the following code:

```
import json
from IPython.display import Image
from py2cytoscape.util import from_networkx
p53_interactions = from_networkx(graph)
p53_net = requests.post(server + '/networks',
    data=json.dumps(p53_interactions),
    headers={'Content-Type': 'application/json'})
net_id = p53_net.json()['networkSUID']
requests.get('%s/apply/layouts/circular/%d' % (server,
    net_id))
Image('%s/networks/%d/views/first.png' % (server, net_id))
```

- ❑ Here, we will use a `py2cytoscape` utility function to convert the `NetworkX` representation to a JSON version, that can be sent to Cytoscape.

- ❑ Remember to have Cytoscape running, and the `cyREST` plugin installed. We will use the default network interface of `cyREST` here.

- ❑ We will create our graph (networks service with the HTTP POST method), specify a circular layout, and draw the image that Cystoscape renders, as shown in the following figure:

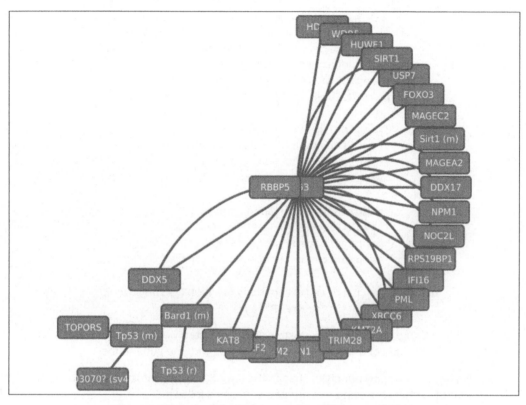

Figure 4: A first approach with Cytoscape for a p53 query on UniProt

6. Next, we want a different style for our plot; that is, we want the gene name (currently, it has the name and the species indicator, if not human) in the node. To color each node by species, we define a style as follows:

```
ustyle = {
    'title': 'UniProt style',
    'mappings': [
        {'mappingType': 'discrete',
         'map': [
                {'key': 'human', 'value': '#00FF00'},
                {'key': 'rat', 'value': '#FF00FF'},
                {'key': 'mouse', 'value': '#00FFFF'}],
         'visualProperty': 'NODE_FILL_COLOR',
         'mappingColumnType': 'String',
         'mappingColumn': 'species'},
        {
            'mappingType': 'passthrough',
            'visualProperty': 'NODE_LABEL',
            'mappingColumnType': 'String',
            'mappingColumn': 'gene'},
        {
            'mappingType': 'passthrough',
            'visualProperty': 'EDGE_TOOLTIP',
            'mappingColumnType': 'String',
            'mappingColumn': 'interaction'
        }],
    'defaults': [ {"visualProperty": "NODE_FILL_COLOR",
            "value": "#FFFFFF"}]}
```

❑ There are quite a few existing styles, but here we will color our nodes as a function of the species, labeling a node with the gene name.

7. Finally, we will apply our style, change the layout (do not confuse the graph layout with the graph style), and plot our new version as follows:

```
res = requests.post(server + "/styles",
data=json.dumps(ustyle),
                                headers={'Content-Type':
                                'application/json'})

requests.get('%s/apply/layouts/force-directed/%d' %
     (server, net_id))
res = requests.get('%s/apply/styles/UniProt style/%d' %
                                (server, net_id),
                                headers={'Content-Type':
'application/json'})
Image('%s/networks/%s/views/first.png' % (server, net_id))
```

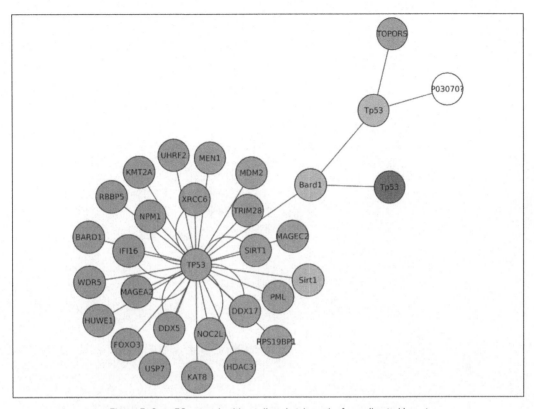

Figure 5: Our p53 network with a tailored style and a force-directed layout

There's more...

The preceding example was designed to exercise more than the REST interface to Cytoscape; we interfaced with the PSICQUIC server and used the NetworkX graph library, functionality than can be useful without Cytoscape. To start with, there is plenty of documentation for NetworkX and Cytoscape available on their web pages.

The **Kyoto Encyclopedia of Genes and Genomes** (**KEGG**) is one of the most useful resources to use with Cytoscape; as this is very well documented, we opted for a less common example based on UniProt. A fantastic documentation to interact with Cytoscape via REST interfaces using IPython, including a KEGG example, is available at http://nbviewer.ipython.org/github/idekerlab/cy-rest-python/blob/develop/index.ipynb.

If drawing KEGG pathways is all that you need, there is a much lighter solution than having to install and use Cytoscape, which is Biopython. You can find the KEGG module documentation at http://nbviewer.ipython.org/github/widdowquinn/notebooks/blob/master/Biopython_KGML_intro.ipynb.

9
Python for Big Genomics Datasets

In this chapter, we will cover the following recipes:

- ▶ Setting the stage for high-performance computing
- ▶ Designing a poor human concurrent executor
- ▶ Performing parallel computing with IPython
- ▶ Computing the median in a large dataset
- ▶ Optimizing code with Cython and Numba
- ▶ Programming with laziness
- ▶ Thinking with generators

Introduction

In this final chapter, we will discuss high-performance computing techniques for large computational biology datasets. We will talk about code parallelization, running software in clusters, code optimization, and advanced functional programming techniques. We will try to avoid tying any solution to a specific proprietary technology (for example, Amazon EC2) and design code that can be applicable in a wide range of scenarios.

The whole topic of persistence is mostly left out of this chapter, although we do make some minor considerations on the IO performance. There is no single good solution for persistence in computational biology; you will probably use SQL datasets for some limited applications. Most of your files will be BAM- or VCF-formatted, and you will probably use a lot of text files too. You may also want to consider NoSQL databases in some instances. Having said that, if you have not checked HDF5, you may want to have a look at it, especially as there are quite a few Python implementations available. Many topics of the following recipes deserve a whole book in their own right. The objective here is not to be exhaustive, but to give you a taste of the possibilities available. You are strongly encouraged to research further if you find any of the following topics interesting.

Setting the stage for high-performance computing

In this recipe, we will prepare you so that you can perform computing with multiple cores, in clusters, and in MapReduce frameworks. We will use a simple example, where we compute the **minimum allele frequency** (**MAF**) of loci across the human genome using the TSI ("Toscani in Italy") HapMap population. Refer to *Chapter 6, Phylogenetics*, for details on the HapMap data.

We will perform two different kinds of tasks here. First is preparing the data, another is structuring computations as if we were using the parallel computing framework. The sequential execution is a safe and predictable environment to introduce parallel programming concepts, even if we still do not do actual concurrent execution in this recipe. We will use this recipe to also introduce some pitfalls with big data processing and a few basic functional programming techniques.

Getting ready

Here, we will need the data we used in the *Managing datasets with PLINK* recipe in *Chapter 6, Phylogenetics*. These files are specified in our dataset list at `https://github.com/tiagoantao/bioinf-python/blob/master/notebooks/Datasets.ipynb` (`hapmap.map.bz2`, `hapmap.ped.bz2`, and `relationships.txt`). This recipe will also require PLINK.

You must decompress the following files:

- `bunzip2 hapmap3_r2_b36_fwd.consensus.qc.poly.map.bz2`
- `bunzip2 hapmap3_r2_b36_fwd.consensus.qc.poly.ped.bz2`

As usual, this is available in the `08_Advanced/Intro.ipynb` notebook, where everything has been taken care of.

How to do it...

Take a look at the following steps:

1. First, let's get only the Toscani individuals, as shown in the following code:

```
import os
tsi = open('tsi.ind', 'w')
f = open('relationships_w_pops_121708.txt')

for l in f:
    toks = l.rstrip().split('\t')
    fam_id = toks[0]
    ind_id = toks[1]
    mom = toks[2]
    dad = toks[3]
    pop = toks[-1]
    if pop != 'TSI' or mom != '0' or dad != '0':
        continue
    tsi.write('%s\t%s\n' % (fam_id, ind_id))
f.close()
tsi.close()
os.system('plink --file hapmap3_r2_b36_fwd.consensus.qc.poly
--maf 0.001 --keep tsi.ind --make-bed --out tsi')
```

2. Here, we will just write a file with all Toscani individuals so that we can call PLINK to extract those. We are careful to bring on the polymorphic SNPs for this population (hence, the MAF filter). Let's extract the maximum position per chromosome for future reference as follows:

```
import pickle
from collections import defaultdict
from collections import defaultdict
max_chro_pos = defaultdict(int)
f = open('tsi.bim')
for l in f:
    toks = l.rstrip().split('\t')
    chrom = int(toks[0])
    if chrom > 22:
        continue
    pos = int(toks[3])
    if pos > max_chro_pos[chrom]:
        max_chro_pos[chrom] = pos
f.close()
w = open('max_chro_pos.pickle', 'w')
pickle.dump(max_chro_pos, w)
w.close()
```

- ❏ We get the maximum position per chromosome so that in the future, we can split computing tasks as a function of the chromosome size. A more complete version should also get the starting position of the chromosome. This is because acrocentric chromosomes may not have any SNPs typed at the very beginning. However, this is enough for our purposes.

- ❏ We pickle (that is persist the data structures) the results to disk and use this in the next recipes. Note that this is not a recommendation to use `pickle` as a persistence mechanism for everything. The `pickle` module has security and performance limitations, is Python-specific, and should probably be used only for small datasets in limited situations. On Python 2, you can use `pickle` or `cPickle`, but on Python 3, these modules were merged under `pickle`.

3. We extract the allele frequency of all SNPs in parts (windows), starting by defining a function to traverse the genome as follows:

```
window_size = 2000000
def traverse_genome(traverse_fun, state=None):
    if state is None:
        state = {}
    for chrom, max_pos in max_chro_pos.items():
        num_bin = (max_pos + 1) // window_size
        for my_bin in range(num_bin):
            start_pos = my_bin * window_size + 1  # 1-start
            end_pos = start_pos + window_size
            traverse_fun(state, chrom, start_pos, end_pos)
```

- ❏ First, we define a window size of 2 Mbp. We will use this to split our computations in blocks of 2 Mbp in later recipes, and also as a basis for windowed analysis. Just note that the two (window sizes and computation blocks) should be in general separated concepts.

- ❏ Next, we will define the `traverse_genome` function. In this function, we divide a genome in blocks of 2 Mbp and apply `traverse_fun` to each block. Note the programming pattern here; we will split a task into parts based on the data and apply a function to each.

- ❏ If the computation for each block is independent of other parts of the genome (a pattern that is quite common), then you can start many of these blocks in parallel; this is something that we will perform in the next recipe. In this case, we will allow an optional interblock communication device (a simple dictionary called state).

- ❏ Defining the correct granularity is fundamental (that is, the computation block size), too small and you will start too many processes with all the CPU and IO overhead that that will entail, too large and you will be losing performance or maybe consuming too much memory per process.

- ❑ The block size for this specific task is set too low. Indeed, this example could be done whole genome in a single go. However, obviously, this is a toy example designed to show concepts and still run fast.

4. Let's see this framework in practice; we now compute the MAF in blocks, as shown in the following code:

```
def compute_MAF(state, chrom, start_pos, end_pos):
    os.system('plink --bfile tsi --freq --out tsi-%d-%d --
chr %d --from-bp %d --to-bp %d' %
                                (chrom, start_pos, chrom, start_pos,
                                 end_pos))
    os.remove('tsi-%d-%d.log' % (chrom, start_pos))
traverse_genome(compute_MAF)
```

- ❑ We use PLINK to compute the MAF. Note that you can instruct PLINK to compute just a subset of the genome with `--chro`, `--from-bp` and `--to-bp`. Each block will be outputted to a file called `TSI-<chrom>-<start-pos>.frq`. For example, `TSI-10-12000001.frq` refers the 10th chromosome from `12000001` to `1400000`.

- ❑ Note that the computation of each block is independent of each other, so the blocks could have been started in parallel without much problem.

- ❑ As we will apply this in a sequential framework, this will be quite slow. Remember that this will make sense in performance terms when applied in a parallel framework.

- ❑ Having said that, this code will probably be slow in a parallel execution anyway; note that all blocks will read from the same original PLINK file. The IO contention is a source of lack of performance in parallel processing to the point that running a task sequentially may even be much faster than running it in parallel.

5. Let's gather some statistics (per 2 Mbp block and genome-wide) and compute the number of observations and the mean MAF, as shown in the following code:

```
from collections import defaultdict
def gather_statistics(state, chrom, start_pos, end_pos):
    try:
        f = open('tsi-%d-%d.frq' % (chrom, start_pos))
    except:
        # empty block
        state['block_mafs'][(chrom, start_pos)] = []
        return
    f.readline()
    for cnt, l in enumerate(f):
        toks = [tok for tok in l.rstrip().split(' ') if tok
            != '']
        maf = float(toks[-2])
```

```
                        state['snp_cnt'] += 1
                        state['sum_maf'] += maf
                        state['block_mafs'][(chrom, start_pos)].append(maf)
                f.close()
        stats = {'snp_cnt': 0, 'sum_maf': 0.0, 'block_mafs':
            defaultdict(list)}
        traverse_genome(gather_statistics, state=stats)
        print(stats['snp_cnt'], stats['sum_maf'] / stats['snp_cnt'])
```

- ❑ We now have a shared data structure called `stats` that is passed from block to block. It contains the sum of all MAFs (`sum_maf`) and the number of SNPs observed (`snp_cnt`).

- ❑ `stats` also contains a `block_mafs` dictionary. This dictionary includes the MAF for each and every SNP. Contrary to `sum_maf` and `snp_cnt`, `block_maf` consumes memory in proportion to the number of SNPs genotyped.

- ❑ Finally, we print the number of SNPs (1, 222, and 126) and the mean MAF (0.2316).

6. Let's now perform something apparently benign and compute the median MAF. We will perform this ourselves and not use NumPy here, as shown in the following code:

```
all_mafs = []
for mafs in stats['block_mafs'].values():
    all_mafs.extend(mafs)
#np.median(all_mafs)
all_mafs.sort()
middle = len(all_mafs) // 2
#array of even size
print((all_mafs[middle] + all_mafs[middle + 1]) / 2)
```

- ❑ First, we will create a `all_mafs` list with all the MAFs (remember that these were structured per block in the preceding step).

- ❑ Then, we will sort the MAFs and compute the median MAF (0.2216), all apparently innocuous.

- ❑ The serious problem with this code is that it will require all values in-memory and a sort operation on them. This is quite feasible with 1.2 million floats in modern machines, but it will not scale well. The computation of the median is an example of a case where you need all values in-memory (compare this with the mean where you need only two variables and an updating function). This is not feasible for many more values or if you have to keep track of larger objects than floats. We will revisit this in a separate recipe.

7. Finally, let's compute the number of markers per chromosome as follows:

```
import functools

def collect_mafs(state, chrom, start_pos, end_pos, block_mafs):
    state[chrom] += len(block_mafs[(chrom, start_pos)])

chrom_cnts = defaultdict(int)
traverse_genome(functools.partial(collect_mafs,
                                  block_mafs=stats['block_mafs']),
                                  state=chrom_cnts)
for chrom in range(1, 23):
    print('%2d\t%6d' % (chrom, chrom_cnts[chrom]))
```

1	100780
2	102377
3	84921
4	75819
5	78013
6	82169
7	67218
8	66803
9	56921
10	65166
11	62182
12	60381
13	46354
14	40525
15	37217
16	38443
17	33175
18	36144
19	21749
20	31872
21	16978
22	16919

- In this case, we will define a traverse function that will take the previously computed `mafs` block and simply add per block the number of SNPs observed to a dictionary with the chromosome as a key.

- Note that the `traverse_genome` function requires a `traverse_fun` function with a signature of `state`, `chrom`, `start_pos`, and `end_pos`, but in this case we would like to have an extra parameter (the previously computed block MAFs). This is quite easy with Python because the `functools` module supports partial function application.

Designing a poor human concurrent executor

We will start writing our own parallel executor. This executor will have no external dependencies, will be very light, and will be able to work on multiple core computers. We will also supply a version for clusters. There is no interprocess communication mechanism other than a shared filesystem. We will make a similar data analysis as in the previous recipe, but now in a real concurrent environment.

Getting ready

You should have read and understood the previous recipe. You will at least need to get the HapMap data and run the pickle part from it.

The code for this can be found in the `08_Advanced/Multiprocessing.ipynb` notebook. Also, there is an external file called `get_maf.py`, which is available next to this notebook.

How to do it...

Take a look at the following steps:

1. Let's start with some boilerplate code, loading the largest chromosome position from the previous recipe, and defining the window size at 2 Mbp, as shown in the following code:

    ```python
    from __future__ import division, print_function
    import pickle
    f = open('max_chro_pos.pickle')
    max_chro_pos = pickle.load(f)
    f.close()

    window_size = 2000000
    ```

2. We will define a function to yield all genome blocks in the chromosome, starting position, and end position form:

    ```python
    def get_blocks():
        for chro, max_pos in max_chro_pos.items():
            num_bin = (max_pos + 1) // window_size
            for my_bin in range(num_bin):
                start_pos = my_bin * window_size + 1
                end_pos = start_pos + window_size
                yield chro, start_pos, end_pos
    ```

3. We will now define an executor class that will take care of running the concurrent code. The class will have a constructor and three methods; the first method has the ability to submit a job, as shown in the following code:

```
import multiprocessing
import subprocess
import time

class Local:
    def __init__(self, limit):
        self.limit = limit
        self.cpus = multiprocessing.cpu_count()
        self.running = []

    def submit(self, command, parameters):
        self.wait()
        if hasattr(self, 'out'):
            out = self.out
        else:
            out = '/dev/null'
        if hasattr(self, 'err'):
            err = self.err
        else:
            err = '/dev/null'
        if err == 'stderr':
            errSt = ''
        else:
            errSt = '2> ' + err
        p = subprocess.Popen('%s %s > %s %s' %
                             (command, parameters, out,
                              errSt),
                             shell=True)
        self.running.append(p)
        if hasattr(self, 'out'):
            del self.out
        if hasattr(self, 'err'):
            del self.err
```

❑ The class constructor will determine the number of available cores and initialize the running queue. There is also a limit parameter, which will be discussed later.

❑ We will not delve into the intricacies of how to properly run an external application here because this would be too complicated and of little value. We will use an expedite (although not totally secure) solution to help you run a command through the shell. This approach will allow a user to get the standard output and error channels.

4. The second method of our class waits until there is a running slot available or (alternatively) if the running queue is totally empty. To be clear, this method is still part of the class that we defined in the preceding code:

```python
def wait(self, for_all=False):
    self.clean_done()
    numWaits = 0
    if self.limit > 0 and type(self.limit) == int:
        cond = 'len(self.running) >= self.cpus -
self.limit'
        elif self.limit < 0:
            cond = 'len(self.running) >= - self.limit'
        else:
            cond = 'len(self.running) >= self.cpus * self.limit'
        while eval(cond) or (for_all and len(self.running)
                             > 0):
            time.sleep(1)
            self.clean_done()
            numWaits += 1
```

- ❑ If you pass for_all as True, the code will wait for all running processes to terminate (effectively creating a barrier).

- ❑ If not, it will wait for a slot to become available using self.limit as follows. If limit is an integer bigger than 0, it's the expected number of CPI cores that will not be used, for instance, if there are 32 cores and a limit of 6, the system will try to never ago above 26 running processes. A float between 0 and 1 will be interpreted as the fraction of CPUs to be used, for example, with 32 cores, 0.25 will use at most eight tasks. A negative value will be interpreted as the maximum number of processes that can be executed in parallel, for example, -4 will allocate at most four processes.

5. Finally, we perform a cleanup of completed processes (this being another method in the class) as follows:

```python
def clean_done(self):
    dels = []
    for rIdx, p in enumerate(self.running):
        if p.poll() is not None:
            dels.append(rIdx)
    for del_ in reversed(dels):
        del self.running[del_]
```

- ❑ This code checks all running processes to see whether they are terminated (using the poll() method) and removes them from the running list.

6. Let's now use this code to compute the MAF as follows:

```
import os
executor = Local(limit=-4)
for chrom, start_pos, end_pos in get_blocks():
    executor.submit('plink',
                    '--bfile tsi --freq --out tsi-%d-%d --
chr %d --from-bp %d --to-bp %d' %
                                (chrom, start_pos, chrom, start_pos,
                                end_pos))
executor.wait(for_all=True)
for chrom, start_pos, end_pos in get_blocks():
    os.remove('tsi-%d-%d.log' % (chrom, start_pos))
```

   ❑ We start by creating an executor with at most four concurrent jobs.

   ❑ We iterate over all the blocks and submit jobs for execution that will use PLINK to extract the MAF per block. Note that at 2 Mbp, there are 1425 blocks spanning autosomes and the X chromosome of humans. Therefore, 1425 processes will be created.

   ❑ We will then wait for all processes to terminate and then remove log files as they are not needed.

7. Now, we parse and retrieve the MAF per block, as shown in the following code:

```
for chrom, start_pos, end_pos in get_blocks():
    executor.submit('python', 'get_maf.py %d %d' %
                    (chrom, start_pos))
executor.wait(for_all=True)
```

   ❑ As this framework does not allow us to directly run the Python code (an external process has to be explicitly started), we have to put this code in a separate file.

8. Thus, we will create a simple program called `get_maf.py` to parse the PLINK output and retrieve the MAF, as shown in the following code:

```
import pickle
import sys


def gather_MAFs(chrom, start_pos):
    mafs = []
    try:
        f = open('tsi-%d-%d.frq' % (chrom, start_pos))
        f.readline()
        for cnt, l in enumerate(f):
```

```
                    toks = [tok for tok in l.rstrip().split(' ') if
                        tok != '']
                    maf = float(toks[-2])
                    mafs.append(maf)
                f.close()
            except:
                # might be empty if there are no SNPs
                pass
            w = open('MAF-%d-%d' % (chrom, start_pos), 'w')
            pickle.dump(mafs, w)
            w.close()

chrom = int(sys.argv[1])
start_pos = int(sys.argv[2])
gather_MAFs(chrom, start_pos)
```

- ❑ The `get_maf.py` program gets the chromosome and the starting position from the command line, parses the relevant PLINK file, and outputs the result in a pickle file.

9. Let's now use the MAF data to plot the number of observations per window (of 2 Mbp) for three chromosomes. We do this back in our original file:

```
from collections import defaultdict
import matplotlib.pyplot as plt
plot_chroms = [1, 13, 22]
chrom_values = defaultdict(list)
for chrom, start_pos, end_pos in get_blocks():
    if chrom not in plot_chroms:
        continue
    block_mafs = pickle.load(open('MAF-%d-%d' % (chrom,
                start_pos)))
    chrom_values[chrom].append(((start_pos + end_pos) / 2,
        len(block_mafs)))
fig = plt.figure(figsize=(16, 9))
ax = fig.add_subplot(111)
ax.set_title('Number of observations per 2Mbp window')
ax.set_xlabel('Position')
ax.set_ylabel('Observations')
for chrom in plot_chroms:
    x, y = zip(*chrom_values[chrom])
    ax.plot(x, y, label=str(chrom))
ax.legend()
```

- ❑ We just read the generated pickle files and count the number of observations per block.

❑ Note that for chromosome 1, there will be a lack of SNPs in the center of the chromosome, whereas for 13 and 22, this will be at the beginning. This is because chromosome 1 is metacentric (that is, the centromere is in the middle) and chromosomes 13 and 22 are acrocentric (the centromere is at one of the extremes). The chromatin type of centromeres are more difficult to genotype than elsewhere, hence the lack of SNPs. For more details about this, refer to `http://en.wikipedia.org/wiki/Centromere`.

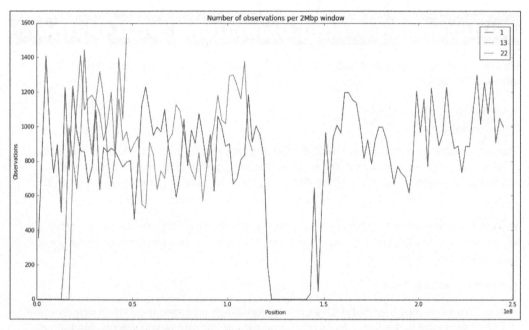

Figure 1: Genotyped SNP density in three different chromosomes for the Toscani population

There's more...

The preceding code gives an example on how you can take full control, in a few lines of code, of a simple concurrent execution. Its value is not only pedagogical; there are many concurrent patterns where splitting the computation into multiple identical blocks with no communication among parallel processes is acceptable.

It's quite trivial to adapt this code to cluster environments. You can find usable code (in need of a cleanup) for LSF, SGE, Torque, and SLURM at `https://github.com/tiagoantao/pygenomics/blob/master/genomics/parallel/executor.py`. The concurrent interface is very similar, with the exception that the code is always nonblocking when submitting tasks.

This is not a beautiful solution, but it has a great advantage over the next recipe; it imposes a few requirements on your basic infrastructure. So, if you are not able to use the next solution (arguably much more elegant), this may be a good enough fallback. Many patterns of usage in computational biology may actually be "embarrassingly parallel" and this approach will actually be enough for many problems. This is the approach I use most of the time; mostly because I have to run the code in extremely heterogeneous environments with varying functionalities and I rarely have the need of complex inter-process communication.

Performing parallel computing with IPython

IPython provides a highly declarative framework for parallel computing. Here, we will take an introductory look at it.

Getting ready

This will require IPython. You have to download and prepare the data, as shown in the first recipe. This recipe will not work in the provided Docker container. It's recommended that you have at least a broader overview of the IPython parallel architecture at `http://ipython.org/ipython-doc/dev/parallel/parallel_intro.html#architecture-overview`.

You will need to start the IPython parallel framework. For this, while inside the directory where you downloaded the data, which is also where you will have to run the recipe code, do in the shell:

```
ipcluster start -n 4
```

This will start the controller with four local engines. Make sure that the Python environment running the cluster is the same as the Python environment, where you will run the recipe.

As usual, this is available in the `08_Advanced/IPythonParallel.ipynb` notebook.

How to do it...

Take a look at the following steps:

1. Let's start with basic imports and loading of the chromosome data prepared in the first recipe:

    ```
    from __future__ import print_function
    import os
    import pickle
    import time
    import numpy as np

    f = open('max_chro_pos.pickle')
    ```

```
max_chro_pos = pickle.load(f)
f.close()

window_size = 2000000

def get_blocks():
    for chro, max_pos in max_chro_pos.items():
        num_bin = (max_pos + 1) // window_size
        for my_bin in range(num_bin):
            start_pos = my_bin * window_size + 1
            end_pos = start_pos + window_size
            yield chro, start_pos, end_pos
```

2. Then, initialize IPython parallel and access the direct interface to the engines:

```
from IPython.parallel import Client
cl = Client()
all_engines = cl[:]
all_engines.execute('import os')
all_engines.execute('import numpy as np')
```

   ❏ If this fails, make sure that `ipcluster` is running with the correct Python version.

   ❏ In this code, we will access all engines (`cl[:]` specifically refers to all the available engines), making sure that all of them import `os` and `numpy`.

3. Let's define a couple of functions to compute and parse MAFs, as shown in the following code:

```
def compute_MAFs():
    for chrom, start_pos, end_pos in my_blocks:
        os.system('plink --bfile tsi --freq --out tsi-%d-%d
--chr %d --from-bp %d --to-bp %d' %
                    (chrom, start_pos, chrom, start_pos,
                     end_pos))
        os.remove('tsi-%d-%d.log' % (chrom, start_pos))


def parse_MAFs(pos):
    chrom, start_pos, end_pos = pos
    mafs = []
    try:
        f = open('tsi-%d-%d.frq' % (chrom, start_pos))
        f.readline()
        for cnt, l in enumerate(f):
            toks = [tok for tok in l.rstrip().split(' ') if
                tok != '']
```

```
                maf = float(toks[-2])
                mafs.append(maf)
            f.close()
        except:
            # might be empty if there are no SNPs
            pass
    return mafs
```

4. The preceding function can be executed locally or remotely, but if we know that we are going to execute them only remotely over a certain set of engines, we can decorate our functions as in these two other functions:

```
@all_engines.parallel(block=True)
def compute_means_with_pos(pos):
    block_mafs = parse_MAFs(pos)
    nobs = len(block_mafs)
    if nobs > 0:
        return np.mean(block_mafs), nobs
    else:
        return 0.0, 0


@all_engines.parallel(block=True)
def compute_means_with_mafs(block_mafs):
    nobs = len(block_mafs)
    if nobs > 0:
        return np.mean(block_mafs), nobs
    else:
        return 0.0, 0
```

  □ If you have never come across decorators, be sure to check, for example,
     http://en.wikibooks.org/wiki/Python_Programming/
     Decorators.

  □ As you will see both kinds of functions (decorated or not), can be remotely
     executed. The decorated ones are defaulted to run remotely.

  □ The decorated functions created earlier perform exactly the same operation
     (computing the mean), but as one gets the block addresses, chromosome,
     the start and end position, the other gets the MAFs per block. The reason for
     these two different versions will become clear soon.

5. Let's now compute the MAFs on the engines using PLINK:

```
all_engines.scatter('my_blocks', list(get_blocks()))
all_engines.apply_sync(compute_MAFs)
```

- ❏ The first line scatters the list of blocks across all four engines. If you reread the preceding `compute_MAF` function, you will see that it takes a global variable (`my_blocks`). This variable is not defined at the start of each available engine, so our client will break it into four pieces (as we have four engines) and distribute `my_blocks`. As such, each engine will do part of the computation.

6. Also, now let's compute the mean MAF per block by passing positions, as shown in the following code:

```
all_engines.push({'parse_MAFs': parse_MAFs})
%timeit compute_means_with_pos.map(get_blocks())
```

- ❏ Note that the `compute_means_with_pos` function requires a `parse_MAFs` function. This function is not available at the start of the engines, so we push it to all the engines. Of course, you can also push "standard" variables to the engine namespace. A pull operation to get a variable from an engine is also available.

7. Let's repeat the computation by first computing all MAFs, getting the values to the client, and then computing their means with the following code:

```
def compute_with_blocks():
    block_mafs = all_engines.map_sync(parse_MAFs,
        get_blocks())
    block_means = compute_means_with_mafs.map(block_mafs)
    return block_means
%timeit compute_with_blocks()
```

- ❏ You may wonder why two versions. Well, note that this second version requires a lot of interprocess communication. It computes the MAFs on all engines, gets them to the client, and passes them back to the engines for the mean computation.

- ❏ With the local executor, all of this is extremely efficient. Also, the timing will be comparable. However, most probably, the second version will be much slower on a cluster (where process communication is very heavy).

8. Finally, let's compute the mean MAF and the number of SNPs. The result will be exactly equal to the previous two recipes:

```
block_means = compute_means_with_pos.map(get_blocks())
sum_maf = 0.0
cnt_maf = 0
for block_maf, block_cnt in block_means:
    sum_maf += block_maf * block_cnt
    cnt_maf += block_cnt
print(cnt_maf, sum_maf / cnt_maf)
```

9. Be aware that there is also an asynchronous interface, as shown in the following code:

```
#blocks are already scattered
async = all_engines.apply_async(compute_MAFs)
import time
#print(async.metadata)
while not async.ready():
    print(len(async), async.progress)
    time.sleep(5)
print('Done')
```

❑ As with asynchronous interfaces, this allows you to start the remote process (or schedule them) and continue the computation.

10. Alternatively, to direct the view of all engines, there is a more declarative interface that will take care of scheduling the tasks for you, as follows:

```
load_balancer = cl.load_balanced_view()
async = load_balancer.map(parse_MAFs,
                          [pos for pos in get_blocks() if
                           pos[0] == 1],
                          block=False, chunksize=3,
                          ordered=True)
while not async.ready():
    print(len(async), async.progress)
    time.sleep(1)
print('Done')
result = async.get()
```

❑ Note that with this interface, we did not specify all engines; there can be 4 or 4000; it's transparent to us.

❑ However, we can give some hints on how to spread the work. For example, the chunksize parameter specifies the size in which the sequences will be broken. In our case, there will be three blocks of 2 Mbp assigned per turn to each engine. By the way, we are only computing MAFs here for chromosome 1.

❑ Note the way to get asynchronous results with the get method.

11. Finally, let's use these results to plot the mean MAF and the 90th percentile for each block across the chromosome 1:

```
import matplotlib.pyplot as plt
%matplotlib inline
fig = plt.figure(figsize=(16, 9))
ax = fig.add_subplot(111)
xs = [x * window_size for x in range(len(result))]
ax.plot(xs, [np.mean(vals) for vals in result])
```

```
ax.plot(xs, [(lambda x : np.percentile(x, 90) if len(x) > 0 else
    None)(vals) for vals in result], 'k-.')
ax.set_xlabel('Chromosome position')
ax.set_xlabel('Heterozygosity')
ax.set_title('Mean and 90th percentile Hz over %d windows
for chromosome 1' % window_size)
```

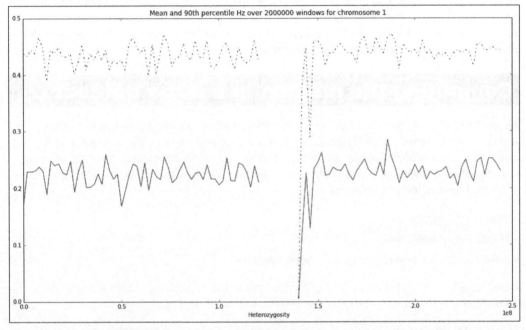

Figure 2: The mean and top ninetieth percentile of MAF across chromosome 1 split into 2 Mbp windows

There's more...

This is a very summarized introduction to the parallel functionality of IPython. There is still much more to say. I would recommend http://ipython.org/ipython-doc/dev/parallel/parallel_intro.html and http://minrk.github.io/scipy-tutorial-2011/ as starting points. Remember that this interface can be used locally and also on a cluster.

While IPython's interface is simple, efficient, and declarative, it may not be for everyone. If you are a cluster user, you probably should check whether the cluster policy easily allows the launching of processes in different nodes, multiple core allocations, and if it is okay to wait for all engines to be up, in order to start a computation. Some clusters are really tailored for batch runs with little synchronicity and communication among nodes (system administrators may not like that a process sits idle waiting for computation, spending a cluster slot). Before considering whether to use IPython parallel on a cluster, I recommend you to perform a test run and see whether the cluster policies and load are aligned with the type of usage that you may want to perform on IPython parallel. For batch and low communication clusters, the previous recipe will actually be more appropriate.

Computing the median in a large dataset

As you have seen in the first recipe, computing the median requires having all the values available. With something like a mean, we just need an accumulator and a counter. The fundamental point of this recipe is to introduce the idea of approximate computing; with big data, it may not always be the best strategy to get the precise value (of course, this should be evaluated on a case-by-case basis).

Getting ready

We will require the first recipe to have been fully run.

Here, we will take two different strategies to compute the median: approximating the data points in a way that allows compression of data and subsampling of data.

As usual, this is available in the `08_Advanced/Median.ipynb` notebook.

How to do it...

Take a look at the following steps:

1. Our first approach will be to use approximations of all values, starting with creating a dictionary. This code should be run where the first recipe was run:

```
from __future__ import division, print_function
import os

from collections import defaultdict
approx = defaultdict(int)
mafs = []
```

```
for fname in os.listdir('.'):
    if not fname.startswith('tsi-') or not
        fname.endswith('.frq'):
        continue
    f = open(fname)
    f.readline()
    for cnt, l in enumerate(f):
        toks = [tok for tok in l.rstrip().split(' ') if tok
            != '']
        maf = float(toks[-2])
        mafs.append(maf)
        approx[maf] += 1
    f.close()
```

- So, instead of having a list of floats, we have a dictionary. In this dictionary, the key is the float and the value is the number of instances that this float occurs as an MAF.

- Note that the MAF varies from 0.0001 to 0.5 and that PLINK rounds to the fourth decimal place anyway, so there is no point in expecting more accuracy. This means that at most, there will be 5000 entries in the dictionary. This is much less than the 1.2 million observations that we found in the first recipe.

2. Let's now check the dictionary:

```
print(len(mafs), type(mafs))
print(len(approx))
```

- So, we have 1,222,126 entries on a list, but only 417 keys in the dictionary. Much better than the worst case of 5000. Why is that? This is because of sampling effects.

- Remember that our sample size is limited (around 100 Toscani individuals). This means that the possible values for the MAF will be constrained by its sample size.

- This is excellent; a worst case scenario would be that the size of an array (remember 1.2 million here, but could be much more with another dataset) is reduced to a maximum of 5000 elements, but is normally much less than this, which is perfectly workable.

3. Before we compute the median, we can plot the distribution of the MAFs just for information purposes, as shown in the following code:

```python
import numpy as np
import matplotlib.pyplot as plt
fig, axs = plt.subplots(2, 2, sharex=True, figsize=(16, 9))

xs, ys = zip(*approx.items())
axs[0, 0].plot(xs, ys, '.')

xs = list(approx.keys())
xs.sort()
ys = [approx[x] for x in xs]
axs[0, 1].bar(xs, ys, 0.005)

def get_bins(my_dict, nbins):
    accumulator = [0] * nbins
    xmin, xmax = xs[0], xs[-1]
    interval = (xmax - xmin) / nbins
    bin_xs = [xmin + i * interval + interval / 2 for i in
        range(nbins)]
    curr_bin = 0
    for x in xs:
        y_cnt = approx[x]
        while curr_bin + 1 != nbins and abs(x -
bin_xs[curr_bin]) > abs(x - bin_xs[curr_bin + 1]):
            curr_bin += 1
        accumulator[curr_bin] += y_cnt
    return bin_xs, accumulator, interval

bin_xs, accumulator, interval = get_bins(approx, 10)
axs[1, 0].bar(np.array(bin_xs) - interval / 2, accumulator,
    0.5 / 10)

bin_xs, accumulator, interval = get_bins(approx, 20)
axs[1, 1].bar(np.array(bin_xs) - interval / 2, accumulator,
    0.5 / 20)
axs[1, 1].set_xlim(0, 0.5)

fig.tight_layout()
```

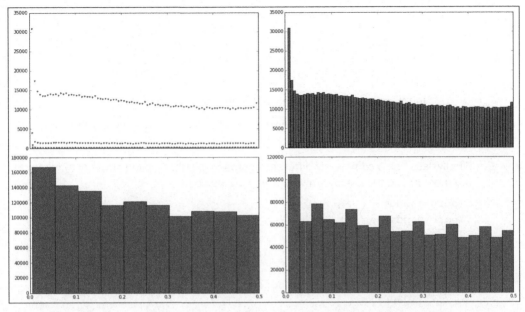

Figure 3: The distribution of MAFs presented as a dot and clustered with different sizes, note that the size if each bar is influenced by the size of the interval that it encompasses and this is related to sampling effects

4. Let's devise an algorithm to compute the median from the preceding dictionary as follows:

```
def compute_median_from_dictionary(my_dict):
    xs = list(my_dict.keys())
    xs.sort()
    x_cnt = [my_dict[x] for x in xs]
    start = 0
    end = len(xs) - 1
    while start != end:
        if start == end - 1 and x_cnt[start] == x_cnt[end]:
            return (xs[start] + xs[end]) / 2
        if x_cnt[start] > x_cnt[end]:
            x_cnt[start] -= x_cnt[end]
            end -= 1
        elif x_cnt[start] < x_cnt[end]:
            x_cnt[end] -= x_cnt[start]
            start += 1
```

```
        else:
            start += 1
            end -= 1
    return xs[start]
print(compute_median_from_dictionary(approx))
```

- ❑ In this case, we will even have the precise value reported.
- ❑ Of course, there is one drawback, that is, we have to write our own function to compute the median.

5. Finally, let's use a completely different approach: a subsampling strategy to compute the median and the maximum with the following code:

```
import random
import pandas as pd
arr = np.ndarray(shape=(5, 6), dtype=float)
samp_sizes = [1, 10, 100, 1000, 10000]
for rep in range(3):
    for si, samp_size in enumerate([1, 10, 100, 1000,
        10000]):
        my_vals = random.sample(mafs, samp_size)
        arr[si, rep] = np.median(my_vals)
        arr[si, rep + 3] = max(my_vals)
df = pd.DataFrame(arr, index=samp_sizes,
                      columns=['Mean #1', 'Mean #2', 'Mean #3',
                      'Max #1', 'Max #2', 'Max #3'])
print(df)   # df
```

	Mean #1	Mean #2	Mean #3	Max #1	Max #2	Max #3
1	0.06818	0.181800	0.07386	0.06818	0.1818	0.07386
10	0.19315	0.170465	0.26135	0.50000	0.4602	0.47160
100	0.23300	0.221600	0.21260	0.49430	0.5000	0.50000
1000	0.22730	0.200000	0.22990	0.50000	0.5000	0.50000
10000	0.22410	0.222550	0.22160	0.50000	0.5000	0.50000

- ❑ The error will depend on the sample size. This example is actually quite benign in the sense that with a very small sample size we get very close to the real value. Not all real-life examples will be like this.

There's more...

Imagine that in a room with 100 people, 98 are of average weight and wealth. However inside, you also have the heaviest human being in the world and Bill Gates. Now, you want to get the mean, median, and maximum of both weight and wealth for the room. Could you get a reasonable approximation if you sample 10 individuals? This would probably work for both medians. For the mean of the weight, the error would probably be still acceptable, but the mean of the wealth would be completely different if Bill Gates would have been in the sample or not. The maximum would probably require all individuals to be sampled, but the order of magnitude of the error would be much higher in the wealth than in the weight.

The preceding example should make clear that your sampling strategy, should you decide to use an algorithm with a subsample of the data, will depend on your dataset and on the metric that you are interested. Bear in mind that the median of the MAF case presented here is actually quite a benevolent example and that most cases will be more difficult than ours.

Optimizing code with Cython and Numba

Here, we will have a short introduction on how to optimize code with Cython and Numba. These are competitive approaches; Cython is a superset of Python that allows you to call C functions and specify C types. Numba is a just-in-time compiler that optimizes the Python code.

As an example, we will reuse the distance recipe from the proteomics chapter. We will compute the distance between all atoms in a PDB file.

Getting ready

Cython normally requires specifying your optimized code in a separate `.pyx` file (Numba is a more declarative solution without this requirement). As IPython provides a magic to hide this, we will use IPython here. However, note that if you are on plain Python, the Cython development will be a bit more cumbersome.

You will need to install Cython and Numba (with `conda`, just perform `conda install cython numba`).

As usual, this is available in the `08_Advanced/Cython_Numba.ipynb` notebook.

How to do it...

Take a look at the following steps:

1. Let's load our PDB structure with the following code:

```
from __future__ import print_function
import math
%load_ext Cython
from Bio import PDB

repository = PDB.PDBList()
parser = PDB.PDBParser()
repository.retrieve_pdb_file('1TUP', pdir='.')
p53_1tup = parser.get_structure('P 53', 'pdb1tup.ent')
```

2. Here is our standard distance function along with its time cost:

```
def get_distance(atoms):
    atoms = list(atoms)   # not great
    natoms = len(atoms)
    for i in range(natoms - 1):
        xi, yi, zi = atoms[i].coord
        for j in range(i + 1, natoms):
            xj, yj, zj = atoms[j].coord
            my_dist = math.sqrt((xi - xj)**2 + (yi - yj)**2
                + (zi - zj)**2)
%timeit get_distance(p53_1tup.get_atoms())
```

  ❑ This will compute the distance between all atoms in the PDB file.

  ❑ The timing will vary from computer to computer, but where this code was tested, it averaged 4 minutes.

3. Let's take a look at the first Cython version, which is nothing more than an attempt to compile this with Cython and see how much time we gain:

```
%%cython
import math
def get_distance_cython_0(atoms):
    atoms = list(atoms)
    natoms = len(atoms)
    for i in range(natoms - 1):
        xi, yi, zi = atoms[i].coord
        for j in range(i + 1, natoms):
            xj, yj, zj = atoms[j].coord
            my_dist = math.sqrt((xi - xj)**2 + (yi - yj)**2
                + (zi - zj)**2)
%timeit get_distance_cython_0(p53_1tup.get_atoms())
```

- ❑ We gained nothing here. Again, around 4 minutes.

- ❑ Indeed, we were not hoping for much. We know that with Cython, the code requires some changes.

4. Let's rewrite the function for Cython and see how much time it takes here:

```
%%cython
cimport cython
from libc.math cimport sqrt, pow

cdef double get_dist_cython(double xi, double yi, double
zi,
                    double xj, double yj, double zj):
    return sqrt(pow(xi - xj, 2) + pow(yi - yj, 2) +
                pow(zi - zj, 2))


def get_distance_cython_1(object atoms):
    natoms = len(atoms)
    cdef double x1, xj, yi, yj, zi, zj
    for i in range(natoms - 1):
        xi, yi, zi = atoms[i]
        for j in range(i + 1, natoms):
            xj, yj, zj = atoms[j]
            my_dist = get_dist_cython(xi, yi, zi, xj, yj,
                zj)
%timeit get_distance_cython_1([atom.coord for atom in
    p53_1tup.get_atoms()])
```

- ❑ So, we took the expensive arithmetic computation that sits in the inner loop and optimized it.

- ❑ We use `libc` (the fast C code) and make sure that Cython has all the necessary typing information.

- ❑ The result, that is, 18 seconds is 12 times better when compared with 4 minutes! This is worthwhile.

- ❑ Note that we only optimized the inner loop, which was highly number crunching. You probably do not want to perform more than this because over optimizing your algorithms tend to make them difficult to read and manage. Also, you get no sizable advantage from optimizing the noninner loop code.

5. We now switch to Numba and use a decorator to create an optimized version of the original function and time it with the following code:

```
from numba import float_
from numba.decorators import jit
get_distance_numba_0 = jit(get_distance)
%timeit get_distance_numba_0(p53_1tup.get_atoms())
```

❑ Again, it's 4 minutes. Here, we had no expectations really because in theory, Numba can optimize lots of code. Maybe, future versions will be able to deal with this code automatically.

6. We can refactor this code to see whether we can get a better result with Numba:

```
@jit
def get_dist_numba(xi, yi, zi, xj, yj, zj):
    return math.sqrt((xi - xj)**2 + (yi - yj)**2 + (zi -
        zj)**2)


def get_distance_numba_1(atoms):
    natoms = len(atoms)
    for i in range(natoms - 1):
        xi, yi, zi = atoms[i]
        for j in range(i + 1, natoms):
            xj, yj, zj = atoms[j]
            my_dist = get_dist_numba(xi, yi, zi, xj, yj,
                zj)
%timeit get_distance_numba_1([atom.coord for atom in
    p53_1tup.get_atoms()])
```

❑ We are now at 38 seconds.

❑ Note that refactoring is also performed as in Cython, but the code is 100 percent Python (whereas the Cython code is close to Python, but not Python).

❑ Also, there was no need to decorate the function extensively. In theory, you could annotate the type of the function and parameters, but Numba does a great job at discovering this for you. Also, you get no improvements (in this case, at least) with more annotations.

There's more...

This is just a very small taste of what both libraries can do. For example, Numba can work with NumPy and generate code for GPUs.

Note that the performance comparison will vary from problem to problem. You can find cases on the Web (where Numba outperformed Cython). Refer to, `https://jakevdp.github.io/blog/2012/08/24/numba-vs-cython/` for examples.

It should be clear that Numba is less intrusive than Cython because you end up with 100 percent Python code (although you may still have to refactor for performance).

Do not over optimize; find the most critical parts of your code and concentrate your efforts there.

Programming with laziness

Lazy evaluation of a data structure delays the computation of values until they are needed. It comes mostly from functional programming languages, but has been increasingly adopted by Python among other popular languages. Indeed, one of the biggest differences between Python 2 and Python 3 is that Python 3 tends to be lazier than Python 2. It turns out that lazy evaluation allows easier analysis of large datasets, generally requiring much less memory and sometimes performs much less computation.

Here, we will take a very simple example from *Chapter 2, Next-generation Sequencing*, we will take two paired-end read files and try to read them simultaneously (as order on both files represents the pair).

Getting ready

We will repeat part of the analysis performed on the FASTQ recipe in *Chapter 2, Next-generation Sequencing*. This will require two FASTQ files (SRR003265_1.filt.fastq. gz and SRR003265_2.filt.fastq.gz) that you can retrieve from https://github. com/tiagoantao/bioinf-python/blob/master/notebooks/Datasets.ipynb:

We will use the timeit magic here, so this code will require IPython. For an alternative and more verbose approach on how to explicitly use the timeit module on standard Python, refer to the *Computing distances on a PDB file* recipe in *Chapter 7, Using the Protein Data Bank*.

As usual, this is available in the 08_Advanced/Lazy.ipynb notebook.

How to do it...

Take a look at the following steps:

1. To understand the importance of lazy execution with big data, let's take a motivational example based on reading pair-end files. Do not run this on Python 2 because your interpreter will crash and probably become unstable at least for some time:

```python
from __future__ import print_function
import gzip
from Bio import SeqIO
f1 = gzip.open('SRR003265_1.filt.fastq.gz', 'rt')
f2 = gzip.open('SRR003265_2.filt.fastq.gz', 'rt')
recs1 = SeqIO.parse(f1, 'fastq')
recs2 = SeqIO.parse(f2, 'fastq')
cnt = 0
for rec1, rec2 in zip(recs1, recs2):
    cnt +=1
print('Number of pairs: %d' % cnt)
```

❑ The problem with this code on Python 2 is that the `zip` function is eager and will try to generate the complete list, thus reading (or trying to and failing spectacularly) both files in-memory. Python 3 is lazy. It will generate two records at a time for every time that there is an for loop iteration. Eventually, the garbage collector will take care of cleaning up the memory. Python 3's memory footprint is negligible here.

❑ This problem can be solved on Python 2; we will see this very soon.

❑ Note that this code relies on the fact that Biopython's parser also returns an iterator, where it will perform an in-memory load of all the files and the problem would still exist. Thus, if you have lazy iterators, it's normally safe to chain them in a pipeline as memory and CPU will be used on need-to-use basis. A chain that includes an eager element may require some care or even rewriting.

2. Probably, the historical example on Python 2 between eager and lazy evaluation will come from the usage of `range` versus `xrange`, as shown in the following code:

```
print(type(range(100000)))
print(type(xrange(100000)))
%timeit range(100000)
%timeit xrange(100000)
%timeit xrange(100000)[5000]
```

❑ The type of the range will be a list as on Python 2 the range function will create a list. This will require time to create all the elements and also allocate the necessary memory. In the preceding case, 1 million integers will be allocated.

❑ The second line will create an object of the `xrange` type. This object will have a very small memory footprint because no list is created.

❑ In terms of timing, this range will run in milliseconds; the `xrange` function in nanoseconds, approximately four order of magnitude faster with no significant memory allocation. The `xrange` type also allows direct access via indexing with no extra memory allocation and constant time in the same order of magnitude of nanoseconds. Note that you will not have this last luxury with normal iterators.

❑ Python 3 has only a range function, which behaves like the Python 2 `xrange`.

3. One of the biggest differences between Python 2 and Python 3 is that the standard library of version 3 is much lazier. If you execute this on Python 2 and 3, you will have completely different results:

```
print(type(range(10)))
print(type(zip([10])))
print(type(filter(lambda x: x > 10, [10, 11])))
print(type(map(lambda x: x + 1, [10])))
```

   ❏ Python 2 will return all as lists (that is, all values were computed), whereas Python 3 will return iterators.

4. Note that you do not lose any generality with iterators because you can convert these to lists. For example, if you want direct indexing, you can simply perform this on Python 3:

```
big_10 = filter(lambda x: x > 10, [9, 10, 11, 23])
#big_10[1] this would not work in Python 3
big_10_list = list(big_10)  # Unnecessary in Python 2
print(big_10_list[1])  # This works on both
```

5. Although Python 2 built-in functions are mostly eager, the `itertools` module makes available lazy versions of many of them. For example, a version of the FASTQ to process the output of a FASTQ paired sequencing run that works on both versions of Python will be as follows:

```
import sys
if sys.version_info[0] == 2:
    import itertools
    my_zip = itertools.izip
else:
    my_zip = zip
f1 = gzip.open('SRR003265_1.filt.fastq.gz', 'rt')
f2 = gzip.open('SRR003265_2.filt.fastq.gz', 'rt')
recs1 = SeqIO.parse(f1, 'fastq')
recs2 = SeqIO.parse(f2, 'fastq')
cnt = 0
for rec1, rec2 in my_zip(recs1, recs2):
    cnt +=1
print('Number of pairs: %d' % cnt)
```

   ❏ There are a few relevant functions on `itertools`; be sure to check `https://docs.python.org/2/library/itertools.html`.

   ❏ These functions are not available on the Python 3 version of `itertools` because the default built-in functions are lazy.

There's more...

Your function code can be lazy with generator functions; we will address this in the next recipe.

Thinking with generators

Writing generator functions is quite easy, but more importantly, they allow you to write different dialects of code that are more expressive and easier to change. Here, we will compute the GC skew of the first 1000 records of a FASTQ file with and without generators discussed in the preceding recipe. We will then change the code to add a filter (the median nucleotide quality has to be 40 or higher). This allows you to see the extra code writing style that generators allow you in the presence code changes.

Getting ready

You should get the data as in the previous recipe, but in this case, you only need the first file called `SRR003265_1.filt.fastq.gz`.

As usual, this is available in the `08_Advanced/Generators.ipynb` notebook.

How to do it...

Take a look at the following steps:

1. Let's start with the required import code:

   ```
   from __future__ import division, print_function
   import gzip
   import numpy as np
   from Bio import SeqIO, SeqUtils
   from Bio.Alphabet import IUPAC
   ```

2. Then, print the mean GC-skew of the first 1000 records with the following code:

   ```
   f = gzip.open('SRR003265_2.filt.fastq.gz', 'rt')
   recs = SeqIO.parse(f, 'fastq',
   alphabet=IUPAC.unambiguous_dna)
   sum_skews = 0
   for i, rec in enumerate(recs):
       skew = SeqUtils.GC_skew(rec.seq)[0]
       sum_skews += skew
       if i == 1000:
           break
   print (sum_skews / (i + 1))
   ```

3. Now, let's perform the same computation with a generator:

```
def get_gcs(recs):
    for rec in recs:
        yield SeqUtils.GC_skew(rec.seq)[0]


f = gzip.open('SRR003265_2.filt.fastq.gz', 'rt')
recs = SeqIO.parse(f, 'fastq',
alphabet=IUPAC.unambiguous_dna)
sum_skews = 0
for i, skew in enumerate(get_gcs(recs)):
    sum_skews += skew
    if i == 1000:
        break
print(sum_skews / (i + 1))
```

□ In this case, the code is actually slightly bigger. We have extracted the preceding function to compute the GC skew. Note that we can now process all the records in that function as they are being returned one by one in case they are needed (indeed, we only need to get the first 1000 records).

4. Let's now add a filter and ignore all records with a median PHRED score that is less than 40. This is the nongenerator version:

```
f = gzip.open('SRR003265_2.filt.fastq.gz', 'rt')
recs = SeqIO.parse(f, 'fastq',
alphabet=IUPAC.unambiguous_dna)
i = 0
sum_skews = 0
for rec in recs:
    if np.median(rec.letter_annotations['phred_quality']) < \
        40:
        continue
    skew = SeqUtils.GC_skew(rec.seq)[0]
    sum_skews += skew
    if i == 1000:
        break
    i += 1
print(sum_skews / (i + 1))
```

□ Note that the logic sits in the main loop. From a code design perspective, this means that you have to tweak the main loop of your code.

□ Interestingly, we now cannot use enumerate anymore to count the number of records because the filtering process requires us to ignore part of the results. So, if we had forgotten to change it, you would have a bug.

5. Let's now change the code of the generator version:

```python
def get_gcs(recs):
    for rec in recs:
        yield SeqUtils.GC_skew(rec.seq)[0]


def filter_quality(recs):
    for rec in recs:
        if np.median(rec.letter_annotations['phred_quality']) >= \
            40:
            yield rec


f = gzip.open('SRR003265_2.filt.fastq.gz', 'rt')
recs = SeqIO.parse(f, 'fastq', alphabet=IUPAC.unambiguous_dna)
sum_skews = 0
for i, skew in enumerate(get_gcs(filter_quality(recs))):
    sum_skews += skew
    if i == 1000:
        break
print (sum_skews / (i + 1))
```

❑ We add a new function called `filter_quality`. The old `get_gcs` function is the same.

❑ We chain `filter_quality` with `get_gcs` in the main for loop and do no more changes. This is possible because the cost of calling the generator is very low as it is lazy. Now, imagine that you need to chain any other operations to this; which code seems more amenable to change without introducing bugs?

See also

▸ To take a look at generator expressions at `http://www.diveintopython3.net/generators.html`

▸ Finally, the amazing *Generator Tricks for System Programmers tutorial* from David Beazley at `http://www.dabeaz.com/generators/`

Index

A

Admixture
population structure,
investigating with 118-123
URL 118
aligned sequences
comparing 164-169
alignment data
working with 37-43
AmiGO
URL 88
Anaconda
distribution, URL 5
URL 2, 7
used, for installing software 2-7
animation
with PyMol 212-220
annotations
used, for extracting genes
from reference 76-79
Arlequin
about 169
URL 169
arXiv
URL 9

B

Bioconductor
documentation, URL 16
URL 15
Bio.PDB
about 192
using 193-196
Bio.Phylo 180

Bio.PopGen
used, for exploring dataset 101-106
Biopython
about 155
coalescent, simulating with 149-153
SeqIO page, URL 28, 30
tutorial, URL 24
URL 28, 59, 153
used, for parsing mmCIF files 220, 221
biostars
URL 24, 43
boot2docker
URL 7
Burrows-Wheeler Aligner (BWA)
URL 43
bzip2
URL 92

C

CDSs
centromeres
URL 259
Cholroquine Resistance Transporter (CRT) 21
coalescent
simulating, with Biopython 149-153
simulating, with fastsimcoal 149-153
code
optimizing, Cython used 271-274
optimizing, Numba used 271-274
Coding Sequence (CDS)
about 26
URL 79
comprehensions
URL 179
Copy Number Variation (CNVs) 89

CRAN
URL 15
Cython
used, for optimizing code 271-274
Cytoscape
URL 239
used, for plotting protein
interactions 239-244

D

dataset
exploring, with Bio.PopGen 101-106
managing, with PLINK 91-96
decorators
URL 262
demographic
complex demographic scenarios,
modeling 143-149
DendroPy 155
Docker
URL 2, 7
used, for installing software 7-9

E

Ebola dataset
preparing 156-162
references 162
Ebola virus
about 156
URL 156
EIGENSOFT
URL 113
Ensembl
gene ontology information, retrieving 83-88
URL 24, 79
used, for finding orthologues 80-83
ETE
about 162
URL 162
executor
URL 259
exons
URL 79
extensions 18

F

FASTQ format
URL 36
fastsimcoal
coalescent, simulating with 149-153
URL 150, 153
fastStructure 118
First-In First-Out (FIFO) 177
Flybase
URL 68
forward-time simulations 126-131
F-statistics
computing 107-112
URL 113

G

GATK 47
GBIF
about 223
accessing 224-229
datasets, geo-referencing 230-236
REST API, URL 229
URL 224
GenBank
accessing 20-24
GeneAlEx
URL 153
gene ontology (GO)
about 191
information, retrieving from
Ensembl 83-88
URL 88
Genepop format
about 97-100
URL 101, 107
generators
about 278-280
expressions, URL 280
genes
data, aligning 162-164
extracting from reference,
annotations used 76-79
genomes
1000 genomes project, URL 9, 37

accessibility 47, 51-59
annotations, traversing 73-76
browser, URL 83
high-quality reference genomes 62-67
URL 68, 83
genomics
data, aligning 162-164
data results FAQ, URL 43
geometric operations
performing 205-208
Gephi
URL 185
GFF spec
URL 76
gffutils
URL 76
ggplot
URL 15
Git
URL 3
Global Biodiversity Information Facility. *See*
GBIF
Global Interpreter Lock (GIL)
URL 138
graph drawing
URL 185
graphviz 179
grep tool 194

H

HapMap
URL 91
hydrogen detection
URL 208

I

IPython
parallel architecture, URL 260
parallel functionality, URL 265
used, for performing parallel
computing 260-264
used, for performing R magic 16-18
IPython magics
URL 18

IPython Notebook
URL 7
itertools
URL 277

K

KEGG (Kyoto Encyclopedia of Genes and
Genomes)
URL 245

L

Last-In First-Out (LIFO) 177
laziness
programming with 275-278
Linkage Disequilibrium
URL 97

M

MAFFT
about 162
URL 162
Map Services
URL 230
median
in large dataset, approximating 266-271
minimum allele frequency (MAF) 248
mitochondrial matrix
URL 87
mmCIF files
format, URL 197
parsing, Biopython used 220, 221
molecular distances
computing, on PDB file 201-204
molecular-interaction databases
accessing, with PSIQUIC 236-239
multiple databases
protein, finding 188-192
MUSCLE
about 162
URL 162
MySQL tables
URL 79

N

National Center for Biotechnology Information (NCBI)
about 20
databases, URL 24
data, fetching 21-24
data, searching 21-24
URL 21
NetworkX graph processing library
URL 239
Next-generation Sequencing (NGS)
about 19
URL 16
notebook
URL 73
Numba
URL 274
used, for optimizing code 271-274

O

OpenStreetMap
URL 230, 236
orthologues
finding, Emsembl REST API used 80-83

P

Panther
URL 88
parallel computing
performing, with IPython 260-266
PCA
about 113, 114
URL 117, 118
PDB file
format, URL 208
information, extracting 197-200
molecular distances, computing 201-204
PDB parser
implementing 208-212
P. falciparum genome
URL 62
Phred quality score 30
phylogenetic trees
about 155

reconstructing 170-174
rooted trees 174
unrooted trees 174
visualizing 179-185
Picard
URL 43
Pillow
URL 230
PlasmoDB
URL 68
PLINK
datasets, managing with 91-97
URL 90
poor human concurrent executor
designing 254-260
population structure
investigating, with Admixture 118-123
simulating, island model used 138-143
simulating, stepping-stone
model used 138-143
URL 118
Principal Components Analysis. *See* **PCA**
Project Jupyter 4
protein
finding, in multiple databases 188-192
Protein Data Bank 187
protein interactions
plotting, Cytoscape used 239-245
proteomics 187
PSIQUIC
URL 236
used, for accessing molecular-interaction
databases 236-239
pygenomics
URL 97
pygraphviz
about 179
URL 179
PyMol
URL 213, 220
used, for animation 212-220
PyProt
URL 197
Python distribution
URL 7
Python library
URL 222

Python software
list 4

R

R
interfacing with, rpy2 used 9-15
magic, performing with IPython 16
URL 6
RAxML
about 170
URL 170
recursive programming
with trees 174-178
reference
genes extracting from,
annotations used 76-79
low-quality reference genomes 68-73
RepeatMasker
URL 73
rpy2
used, for interfacing with R 9-15
rpy library documentation
URL 16

S

SAM/BAM format
URL 43
seaborn
URL 4
selection
simulating 132-136
SEQanswers
URL 43
sequence analysis
performing 25-28
sequence formats
working with 28-37
simcoal
URL 153
simulation
coalescent, with Biopython 149-153
coalescent, with fastsimcoal 149-153
forward-time simulations 126-131

population structure,
island model used 138-143
population structure, stepping-stone model
used 138-143
selection 132-138
simuPOP
about 126
URL 131
Single-nucleotide Polymorphisms (SNPs) 23
SNP data
filtering 47-54
SnpEff
URL 48
stage
setting, for high-performance
computing 248-253

T

tiling coordinates
URL 236
TP53 protein
URL 192
trees
recursive programming 174-178
URL 178
TrimAl
about 162
URL 162

U

UCSC Genome Bioinformatics
URL 24
UniProt's REST interface
URL 192

V

variant call format (VCF)
data, analyzing 44-46
URL 47
VectorBase
URL 68
virtualenv
URL 6

W

well-known text
  URL 236
Whole Genome Sequencing (WGS) 19

X

Xcode
  URL 4

Thank you for buying
Bioinformatics with Python Cookbook

About Packt Publishing

Packt, pronounced 'packed', published its first book, *Mastering phpMyAdmin for Effective MySQL Management*, in April 2004, and subsequently continued to specialize in publishing highly focused books on specific technologies and solutions.

Our books and publications share the experiences of your fellow IT professionals in adapting and customizing today's systems, applications, and frameworks. Our solution-based books give you the knowledge and power to customize the software and technologies you're using to get the job done. Packt books are more specific and less general than the IT books you have seen in the past. Our unique business model allows us to bring you more focused information, giving you more of what you need to know, and less of what you don't.

Packt is a modern yet unique publishing company that focuses on producing quality, cutting-edge books for communities of developers, administrators, and newbies alike. For more information, please visit our website at www.packtpub.com.

About Packt Open Source

In 2010, Packt launched two new brands, Packt Open Source and Packt Enterprise, in order to continue its focus on specialization. This book is part of the Packt Open Source brand, home to books published on software built around open source licenses, and offering information to anybody from advanced developers to budding web designers. The Open Source brand also runs Packt's Open Source Royalty Scheme, by which Packt gives a royalty to each open source project about whose software a book is sold.

Writing for Packt

We welcome all inquiries from people who are interested in authoring. Book proposals should be sent to author@packtpub.com. If your book idea is still at an early stage and you would like to discuss it first before writing a formal book proposal, then please contact us; one of our commissioning editors will get in touch with you.

We're not just looking for published authors; if you have strong technical skills but no writing experience, our experienced editors can help you develop a writing career, or simply get some additional reward for your expertise.

Bioinformatics with R Cookbook

ISBN: 978-1-78328-313-2   Paperback: 340 pages

Over 90 practical recipes for computational biologists to model and handle real-life data using R

1. Use the existing R-packages to handle biological data.

2. Represent biological data with attractive visualizations.

3. An easy-to-follow guide to handle real-life problems in Bioinformatics like Next-generation Sequencing and Microarray Analysis.

Learning Python Data Visualization

ISBN: 978-1-78355-333-4   Paperback: 212 pages

Master how to build dynamic HTML5-ready SVG charts using Python and the pygal library

1. A practical guide that helps you break into the world of data visualization with Python.

2. Understand the fundamentals of building charts in Python.

3. Packed with easy-to-understand tutorials for developers who are new to Python or charting in Python.

Please check **www.PacktPub.com** for information on our titles

Python Network Programming Cookbook

ISBN: 978-1-84951-346-3   Paperback: 234 pages

Over 70 detailed recipes to develop practical solutions for a wide range of real-world network programming tasks

1. Demonstrates how to write various besopke client/server networking applications using standard and popular third-party Python libraries.

2. Learn how to develop client programs for networking protocols such as HTTP/HTTPS, SMTP, POP3, FTP, CGI, XML-RPC, SOAP and REST.

3. Provides practical, hands-on recipes combined with short and concise explanations on code snippets.

Python Data Visualization Cookbook

ISBN: 978-1-78216-336-7   Paperback: 280 pages

Over 60 recipes that will enable you to learn how to create attractive visualizations using Python's most popular libraries

1. Learn how to set up an optimal Python environment for data visualization.

2. Understand the topics such as importing data for visualization and formatting data for visualization.

3. Understand the underlying data and how to use the right visualizations.

Made in the USA
Coppell, TX
30 May 2020